Nature's Edge

SUNY series in Environmental Philosophy and Ethics
J. Baird Callicott and John van Buren, editors

Nature's Edge

Boundary Explorations in Ecological Theory and Practice

Edited by
Charles S. Brown and
Ted Toadvine

State University of New York Press

Published by
State University of New York Press, Albany

© 2007 State University of New York

All rights reserved

Printed in the United States of America

No part of this book may be used or reproduced in any manner whatsoever without written permission. No part of this book may be stored in a retrieval system or transmitted in any form or by any means including electronic, electrostatic, magnetic tape, mechanical, photocopying, recording, or otherwise without the prior permission in writing of the publisher.

For information, contact State University of New York Press, Albany, NY
www.sunypress.edu

Production by Kelli W. LeRoux
Marketing by Anne M. Valentine

Library of Congress Cataloging-in-Publication Data

Nature's edge : boundary explorations in ecological theory and practice / edited by Charles S. Brown, Ted Toadvine.
    p. cm. — (SUNY series in environmental philosophy and ethics)
  Includes bibliographical references and index.
  ISBN-13: 978-0-7914-7121-0 (hardcover : alk. paper)
  ISBN-13: 978-0-7914-7122-7 (pbk. : alk. paper)
  1. Human ecology—Philosophy. 2. Boundaries—Philosophy. 3. Environmental protection. 4. Environmental ethics. I. Brown, Charles S., 1950– II. Toadvine, Ted, 1968–.

GF21.N37 2007
304.2—dc22

200623731

Contents

Acknowledgments   vii

*An Introduction to the Problem of Boundaries in
Ecological Theory and Practice*
Charles S. Brown   ix

Part I: The Human/Nature Divide and the Nature of Boundaries

Chapter 1:   *Boundaries and Darwin: Bridging the Great Divide*
Max Oelschlaeger   3

Chapter 2:   *Lamarck Redux: Temporal Scale as the Key to the Boundary Between the Human and Natural Worlds*
J. Baird Callicott   19

Chapter 3:   *The Ethical Boundaries of Animal Biotechnology: Descartes, Spinoza, and Darwin*
Strachan Donnelley   41

Chapter 4:   *Cutting Nature at the Seams: Beyond Species Boundaries in a World of Diversity*
Jon Jensen   61

Chapter 5:   *Respect for Experience as a Way into the Problem of Moral Boundaries*
Charles S. Brown   83

Chapter 6:   *Boundarylessness: Introducing a Systems Heuristic for Conceptualizing Complexity*
Beth Dempster   93

Part II: Community, Values, and Sustainability

Chapter 7:   *Boundaries on the Edge*
Irene J. Klaver   113

Chapter 8:	Remapping Land Use: Remote Sensing, Institutional Approaches, and Landscape Boundaries Firooza Pavri	133
Chapter 9:	Boundaries, Communities, and Politics Anna L. Peterson	145
Chapter 10:	The Moral Economy and Politics of Water in the Arid American West T. Clay Arnold	161
Chapter 11:	The Need for a Taxonomy of Boundaries Wes Jackson and Jerry Glover	177
Chapter 12:	How to do Things with Food: A Plea for Multiple Ontologies Bruce Hirsch	195
Chapter 13:	Culture and Cultivation: Prolegomena to a Philosophy of Agriculture Ted Toadvine	207

Notes on Contributors 223

Index 227

Acknowledgments

The chapters in this book were originally written for a conference in Matfield Green, Kansas, which was hosted by The Land Institute and Emporia State University. We owe a great deal of appreciation to Wes Jackson, the staff at the Land Institute, and the townsfolk of Matfield Green for all their generosity and hospitality during the conference. We appreciate and fondly remember undergraduates Matt Lexow, Taylor Hammer, and Nathan Hall who provided logistic support for this meeting. We thank the Social Sciences Department at Emporia State University for its encouragement and moral support. We especially thank Jacque Fehr for her editorial assistance, typing, and sound judgment in the preparation of this text. Our thanks go to Dianne Brown for the artwork on the cover. We are greatly indebted to the contributors whose willingness to revise their chapters to meet the needs of this volume has greatly facilitated this project. We also wish to thank the anonymous reviewers whose efforts have made this text better. Lastly, it has been a delight to work with Jane Bunker of the State University of New York Press whose encouragement and support of this project are greatly appreciated.

An Introduction to the Problem of Boundaries in Ecological Theory and Practice

Charles S. Brown

Divisions, boundaries, thresholds, and limits structure our lives, our concerns, and the world around us. Some of these boundaries are natural divisions: the shell of an egg, our own skin, or the ecotone where the forest ends and the prairie begins. Other boundaries arise from our actions and evaluations, from our investment of concern in some things rather than others. Along with the structures and boundaries of the natural world, the divisions of time and place, self and other, and of good and evil form the context of our actions, our decisions, and our lives. Although all disciplines of human knowledge and practice—and, arguably, all living things—draw and maintain boundaries, no discipline has yet developed that studies the nature of boundaries themselves. What is a boundary? What circumstances and context allow boundaries to form, to be put into play, to be defined, and to be maintained? What different types of boundaries should be distinguished, and how are they similar or different? According to what criteria might boundaries be evaluated and perhaps redrawn?

Of the many questions related to boundaries, two confront environmental activists and theoreticians more directly than the rest: (1) how do boundaries originate and function, especially the boundary between humans and nature, and (2) what is the role of boundaries in establishing a common framework for theory and practice? Attempts to deal constructively with these issues often suffer from a limiting disciplinary approach. By defining our problems as either economic or biological, political or philosophical, we reproduce the structure of the academy, but fail to appreciate the kind of essential interconnections that ecological thinking in particular has emphasized. Real, sustainable solutions to our environmental problems are far more likely to emerge from a truly interdisciplinary approach to core issues, one that remains true to the complex nature of the problems themselves.

Problems of boundary formation and negotiation recur at all levels, and coming to an understanding of the nature and types of boundaries poses a truly interdisciplinary challenge to environmental thinkers.

As the first sustained investigation of the problem of boundaries, this volume lays the foundation for a new, transdisciplinary field of study that involves the natural and social sciences as well as the humanities and cuts across the traditional division of theory from practice. The authors contribute a range of perspectives and approaches to the problem of developing a "taxonomy" of boundaries, a systematic understanding of the nature and types of boundaries relevant to environmental thinking. Although philosophical concerns with boundary questions are present in ancient and contemporary thought, this book explores a new focus that has received interest at the edge of many disciplines but little multidisciplinary treatment as a field of its own. The chapters in this volume make a case for this new approach—a case relevant to philosophy, ecology, geography, sociology, political studies, and a variety of related fields. By forging new ways of thinking about and working with boundaries within the context of an interdisciplinary dialogue, this investigation pursues the ultimate goal of developing practical and sustainable responses to our environmental problems.

The chapters in Part I address fundamental philosophical questions concerning the boundary between nature and culture, the nature of boundaries, and the question of whether boundaries exist apart from human interests and conceptualizations. These chapters feature the fundamental relationship between humans and nature: Is there a boundary to be drawn between the human and the natural and, if so, how should it be conceived? To what extent is the "environmental crisis" a consequence of our "natural" tendencies as humans, and how does this influence our strategies for responding to environmental problems? The chapters in Part II examine the formation, discovery, maintenance, and measurement of boundaries in relation to the various projects of sustainability: What roles do circumstances and context play in the institution, negotiation, and maintenance of boundaries? According to what criteria might boundaries be evaluated and perhaps redrawn? While the chapters in Part I are principally concerned with theoretical issues present in ecological thinking, those in Part II are mostly concerned with problems in ecological practice.

The first three chapters directly address what may be the most basic boundary problem of all: how to understand properly the human/nature divide. These chapters focus on the origin of the human/nature boundary, its philosophical justifications, and the implications of this boundary for environmental theory and practice. In the opening chapter, "Boundaries and Darwin: Bringing the Great Divide," Max Oelschlaeger claims that the Great Divide—Western culture's dominant narrative placing humanity and culture separate from and superior to nature has become pathological and is currently leading us to disaster. Because of the Great Divide we conceptualize and address environmental problems in the wrong ways. We have spent billions on environmental litigation under the Superfund Law while only a fraction of that on pollution prevention. Several billion dollars annually are spent on fire suppression in fire-adapted public forestlands, while only a few tens of millions per annum go toward restoration projects that would re-establish fire as a natural disturbance regime.

Oelschlaeger writes that changing a story that "constitutes cultural bedrock" will not be easy as it calls for a revolution in human self-understanding. He claims that even though Darwin's work has shown that the Great Divide is scientifically untenable and ethically bankrupt it has nevertheless, "through the history of effects, become a defining characteristic of the human condition." By recognizing the Great Divide as an artifact of language that is maintained by a categorical separation of culture and nature, reinforced by a value-hierarchal understanding of the pair, and metaphysically grounded by mind-body dualism, Oelschlaeger argues that we are able to tease this old story apart and reweave it with new themes, ultimately constructing a new legitimating narrative that begins with Darwin and ends in a transformation of human self-understanding.

In the second chapter, "Lamarck Redux: Temporal Scale as the Key to the Boundary Between the Human and Natural Worlds," J. Baird Callicott argues that the division between nature and culture may be traced to differences between the temporal scales and cultural and biological evolution. Cultural evolution proceeds at a radically different pace from the biological evolution of species; it is much faster than the evolution fueled by natural selection because it is Lamarckian, not Darwinian. What renders strip mines, clear-cuts, and beach developments unnatural is not that they are anthropogenic—for, biologically speaking, Homo sapiens is as natural a species as any other—but that they occur at temporal and spatial scales that were unprecedented in nature until nature itself evolved another mode (the Lamarckian mode) of evolution: cultural evolution. This insight allows us to recognize today's mass extinctions as a boundary violation of these differing temporal scales, to establish norms for environmental ethics, and to defend conservation biology's classic norm of naturalness.

In the next chapter, "The Ethical Boundaries of Animal Biotechnology: Descartes, Spinoza, and Darwin," Strachan Donnelley argues that environmental thinking must address the moral significance, integrity, and flourishing of natural communities of organisms and ecosystems. But recent biotechnology, by blurring the boundary between the natural and the artificial, raises significant theoretical problems for any attempt to safeguard what is natural. To address these issues, the roots of our current thinking must be reexamined, and the appropriate starting point is with the traditional debate between Descartes and Spinoza. Donnelley argues that a Spinozistic cosmology, based on the notion of internal relations and a conative conception of existence, has decided philosophical advantages over the Cartesian metaphysics of separate substances that dominates contemporary thought. This alternative metaphysical perspective makes possible ethical judgments that respect the integrity and well-being of ecosystems and animal populations, as well as the naturally evolved, conative capacities and behavior of individual organisms.

The following three chapters focus on the question of what kinds of boundaries are found in nature and which sorts of boundaries are a product of human interests and conceptualizations. In chapter 4, "Cutting Nature at the Seams: Beyond Species Boundaries in a World of Diversity," Jon Jensen critically examines the pivotal role given to the category and concept of species within contemporary biology and environmental policy. Using the cases of wolves in the Northeast and Salmon in the Northwest as his central examples, he focuses primarily on two questions: (1) Is the special role allocated to the category of species consonant with a fully evolutionary perspective? (2) Is a species-based approach sufficient for protecting the full range of biological diversity and the evolutionary and ecological processes on which it depends? Jensen argues that, although species are real, they are not the basal units of either taxonomy or evolutionary theory and are consequently not "special" in the way that much science and conservation policy implies. Rather than seeing this conclusion as a threat to the Endangered Species Act, however, Jensen argues that his approach would strengthen conservation efforts by extending the focus of such efforts up to ecosystems and ecological processes and down to populations and evolutionary units.

In chapter 5, "Respect for Experience as a Way into the Problem of Moral Boundaries," Charles S. Brown argues that the contents of our moral experiences, if studied seriously rather than dismissed as irrational sentiment, offer clues for the development of a moral rationality inherent in our moral intuitions, and that this moral rationality points toward new ways of including the nonhuman within the boundaries of the moral community. According to Brown, our thinking is currently dominated by an instrumental rationality that dismisses moral sentiments as subjective and private, thereby making moral philosophy in general and environmental ethics in particular impossible. The theory of moral rationality that Brown develops provides an alternative to this instrumental thinking, and he explores the consequences of this new rationality for ecological philosophy and environmental ethics.

In the final chapter of Part I, entitled "Boundarylessness: Introducing a Systems Heuristic for Conceptualizing Complexity," Beth Dempster develops a boundaryless system model, which she terms "sympoietic," as an alternative to traditional systems heuristics that emphasize the importance of boundaries. Avoiding the tendencies toward binary opposition and restriction of focus that accompany traditional system heuristics, the sympoietic system is complex, boundaryless, and collectively producing. Dempster contrasts the sympoietic with the autopoietic system proposed by Maturana and Varela while arguing that the sympoietic system is more productive for understanding self-organizing systems like the West coast temperate rainforest and the wasp/orchid symbiosis. Her search for richer metaphors to describe the interconnected and interdependent nature of social-ecological systems lead her to examine Deleuze and Guattari's "decidedly sympoietic" notion of "rhizome."

These chapters are concerned with the nature/culture boundary as well as the boundaries within nature that affect the practical goals of conservation, including the way we construct the boundaries of the moral community, and finally with the prospect of creating a boundaryless system heuristic to better conceptualize complexity. In the next group of chapters, the emphasis on the conceptualization and construction of boundaries that affect ecological practice recedes while the concern with practical and value issues stemming from the interaction of human communities with the natural world comes to the fore. The next four chapters explore how the generation, maintenance, and negotiation of boundaries inform our understanding of the natural world and our place in it. These chapters address the relationship between the natural world and community practices and values, as these practices and values are shaped by religious, economic, scientific, and political considerations. The final three chapters focus on problems of global environmental accounting and food production, with a view toward developing alternative cultural and social practices that support an ecologically sustainable future.

In "Boundaries on the Edge," Irene J. Klaver focuses on the "edge" of boundaries—the functional dynamic of boundaries as places of potential transition, transformation, and translation. She explores this insight through Wittgenstein's notion of understanding as the "seeing of connections" and his emphasis on the importance of intermediate cases in this process. In working out various strategies to develop intermediate cases as boundary processes she shows how boundary objects as diverse as ecotones, watersheds, corridors, the Berlin wall, coyotes, green eyes, and bird migrations facilitate understanding and collaboration across heterogeneous groups. She argues that the oxymoron, as a co-presence of two mutually exclusive meanings, challenges dualistic modes of thinking and is mirrored by the ecotone, the area where two different ecosystems meet. Her focus is not on the boundary as simply a line of division but on the power of boundary as an area of co-constitution.

In "Remapping Land Use: Remote Sensing, Institutional Approaches, and Landscape Boundaries," Firooza Pavri surveys current approaches to conceptualizing landscape boundaries using imaging technology and institutional techniques. She develops an original framework for linking the information generated by both approaches to arrive at a more ecologically and institutionally meaningful understanding of the boundaries that emerge in land use and management. Pavri argues that while satellite imaging data has proven invaluable for resource management that aims to maintain ecological stability and safeguard local livelihoods in sensitive forest areas, such data is of limited value without a firm understanding of the institutional factors affecting forest use, as revealed by the complex socioeconomic interaction patterns of forest use and extraction observed on the ground. Her chapter discusses the promises and challenges of using remote sensing technology to monitor changing land use and land cover patterns in the forest regions of developing countries.

Anna L. Peterson's "Boundaries, Communities, and Politics" examines the concept of community, its political implications, and the role that common frameworks and boundaries play in the constitution of actual communities. To this end, she describes in detail two religiously grounded rural communities: the repopulated communities in northern El Salvador and the Amish and Mennonite communities in the U.S. Midwest. These descriptions concentrate in particular on each community's relation with the natural world and with structural social change, as well as the role religion plays in shaping these relations. On the basis of her descriptions, Peterson explores the similarities and contrasts in how each community constructs and maintains boundaries, when and how these boundaries are crossed, and if and how the communities manage to address urgent political, economic, and environmental problems that they face.

The following chapter, "The Moral Economy and Politics of Water in the Arid American West," by T. Clay Arnold continues the theme of human communities' relations to the natural world by arguing that a proper understanding of the role of water in the arid American West requires recognition of water's value above and beyond its economic utility. He contends instead that water must be understood as a "social good" because it establishes, reproduces, and symbolizes important individual and collective senses of self. As a social good, water carries politically significant normative features that inform westerners' determinations of the legitimacy or illegitimacy of water-related practices and policies. Consequently, the case of water demonstrates the need for a conception of "moral economy" lacking on the current horizon of political theory. Through a detailed analysis of the historical and current treatment of water in western culture and policy, Arnold contrasts the moral economical account of such community practices with standard explanations in terms of elitist, pluralist, institutional, and market culture dynamics and imperatives.

Wes Jackson and Jerry Glover argue in "The Need for a Taxonomy of Boundaries" that we currently lack a formal language that bridges the middle ground between the bookkeepers of business, concerned with profits according to standard economic models, and global accountants, those whose task must be to measure the impact of our extractive economy on the depletion of the earth's natural capital. This formal language must be developed starting from a "taxonomy of boundaries," specifying in particular the gap between our boundaries of consideration—what we consider relevant in making a decision or taking an action—and the boundaries of causation, the effects that our action will actually have over time. For Jackson and Glover, traditional models of accounting fail to address this "middle ground" between our economic system and the global environmental crisis, as they are caught up in a "knowledge-as-adequate" worldview that overestimates our ability to predict the long-term effects of our industrial and agricultural activities. Jackson and Glover put forward the example of egg production to explore what may be learned from this project of middle-ground accounting, comparing the wild, domestic, and industrial egg production systems to discover what boundaries and costs each system masks.

In "How to do Things with Food: A Plea for Multiple Ontologies," Bruce Hirsch explores changes in the foundation world that are opening new directions for philanthropic support of research into ecologically sound agriculture and food production systems. In order to have an effect on environmental policies and agricultural practices, Hirsch notes, it is first necessary to have an adequate conception of social change and its relation to individual values and priorities. Drawing on Heidegger's account of human existence as "being-in-the-world," Hirsch develops an account of our relation with the world that emphasizes the role of our performative and public acts of disclosure—acts by which our identity is formed and the world is disclosed to us from a certain perspective, but of which we are not necessarily self-consciously aware. Hirsch explores ways that philanthropic organizations, by recognizing this disclosive aspect of our relation to the world, can encourage a change in social practices that would, in the long run, shift the human relationship with nature toward a more sustainable pattern.

Ted Toadvine's final chapter, "Culture and Cultivation: Prolegomena to a Philosophy of Agriculture," identifies agriculture as the fundamental boundary or point of transmission between nature and culture. Examining the etymology of the world "cultivation," he discloses a fundamental ambivalence toward agriculture, as the origin of culture that is, simultaneously, excluded from culture. Toadvine seeks the fundamental meaning of the agricultural way of life, first by exploring the symbolism of the seed, which marks an alteration in the human orientation toward temporality and death. Turning then to the work of Joseph Campbell, Gilles Deleuze, and

Félix Guattari, he investigates the notion of animal and plant modes of life, as expressed by the different relations that human societies adopt toward death and the natural world. Drawing on these resources, Toadvine suggests a "rhizomatic" agriculture that might serve as an alternative to our traditional seed-based agriculture (and culture more generally). A rhizomatic approach, he suggests, would be nondualistic and essentially multiple, encouraging both diversity and sustainability.

Part I

The Human/Nature Divide and the Nature of Boundaries

Chapter 1

Boundaries and Darwin: Bridging the Great Divide

Max Oelschlaeger

The introduction to this collection poses questions concerning boundaries between the human estate and, for all practical purposes, everything else. I offer an analysis grounded in a post-Darwinian perspective on humankind as storytelling culture-dwellers, that is, Homo narrans. My thesis is simple. The narratives that presently dominate Western civilization no longer fit with biophysically evolved realities. The heart of the problem is the so-called Great Divide—a boundary that assumes the separation from and dominance over nature by culture. Given that the Great Divide is an artifact of language, there are reasons to believe that the narrative can be changed. But changing a story that constitutes cultural bedrock will not be easy.

Others have touched on these issues in insightful ways. In particular, Marjorie Grene (1986) and Derek Bickerton (1990; 1995) essentially lay out the conceptual figure that I intend to color with evolutionary hues. Bickerton's remarkable argument makes clear that the distinguishing characteristic of humankind is language. And yet language constitutes at one and the same time the wonder and terror of the human predicament. Our linguistically constructed and maintained stories can be pathological. Grene helps us at least glimpse the challenge of changing the dominant story. Any new story must simultaneously facilitate sustainable relations with the biophysical world while maintaining a sense of human distinctiveness.

There are at least three characteristics that define the Great Divide. First is the categorical separation of culture and nature, a conceptual boundary that plays out in a thousand and one different cultural narratives. Even among late Paleolithic and early Neolithic cultures, as well as among the surviving remnants of primary oral people, the separation is evident. For example, totemism is a system of classifying or bounding of the world in which humans share a kinship with animal or plant others yet also retain a distinct identity. Similarly, the myth of the eternal return (its practice still is evident among tribal people of the American Southwest) recognizes a separation between the human world and the natural world, but through performance attempts to ensure the continuing connection between the worlds.

Second, the separation is reinforced through hierarchy, so that culture is over nature, active, on top, in control of a passive and inert nature. Just think of the metaphoric: up, on top, is good. Down, on the bottom, is bad. One is high, the other low, and beneath. One presses down on the other, in a superior or elevated position; the other is weighed upon, in an inferior or depressed condition. Characteristic one, that is, the basic sense of separation, occurs early in human history, even before the advent of the Neolithic. Separation two appears much later, first beginning with agri-cultures, but reaching a fuller development in the 1600 to 1700s. The great apologists of the modern age, Bacon and Descartes, articulate new chapters of the story. Science was narratized as a power through which MAN becomes the master and possessor of nature. MAN acts on nature, and not the other way around, through causal control.

Third, the separation assumes a stronger, metaphysical configuration (better called "a figuration").[1] Culture becomes the locus of human self-identity (person) and collective significance (people). Nature becomes merely the stage on which cultural and personal life plays out. Pushed to a deeper level of boundedness, culture equates with spirit or mind, nature equates with body or matter. Mind-body dualism—for example, as expounded by Descartes—is one consequence. Culture is the native home of psyche, which is the essence of humanity. And nature is the domain of soma, the baser, physical aspects of existence. Culture, as a generation of feminist scholarship makes clear, is construed as the native home of the masculine gender—that is, MAN. MAN carries on the important cultural work, such as science, business, industry, invention, war, and politics. Woman is associated with body and the body of nature. Her work is the household, the preparation of substances to nourish the body, and the reproductive activities of family. "Savages" or "primitives," under this schematism, are human individuals living on the natural side of the divide. (Even Darwin sniffed as he rounded Tierra del Fuego and saw the naked bodies of the indigenes.)

Clearly, boundaries exist. The Great Divide is a deeply entrenched reality. But why? *I believe it is possible to seek the origin of the Great Divide in the procliv-*

ity of humankind to separate itself from nature. But "averted vision," much like that necessary for naked eye astronomy, is required to see it fully.[2] All analogies are subject to disanology, but consider that even under the darkest of night skies the stargazer must avert vision in order to see fully the Seven Sisters. Direct vision diminishes the view, and only the brightest of the stars register on the retina. By averting the eyes, twelve (or more for an experienced viewer) of the stars become visible.

My point: although the Great Divide is virtually a cliché among green writers and academics, its deeper structure (the figuration into which humans are socialized) is concealed, virtually invisible. Think of the 3K cosmic background radiation, the hissing noise first heard in the middle of the twentieth century by scientists at Bell Laboratories. What did it mean? Simply this: The noise was the residual of the Big Bang itself. The analogy I am making is that the Great Divide is to the present human condition and the question of boundaries like the 3K background radiation is to the cosmos and the question of origin. It's the lingering consequence of the so-called explosive encephalization and associated changes that changed our primate ancestors from marginal monkeys to...well...us. I'll come back to "explosive encephalization" and crossing the "monkey/human" boundary in a moment.

So what is averted vision? Nothing new, really. Hans-Georg Gadamer (1988), for example, offers insight. "In our contemporary situation," Gadamer argues, "faced as we are with an increasingly widespread anxiety about human existence as such, the issue is the suspicion ... that if we continue...to turn our earth into one vast factory as we are doing at the moment, then we threaten the conditions of human life in both the biological sense and in the sense of specific human ideals even to the extreme of self-destruction" (491). Could this be, he continues, "because of the baleful influence of language?" (ibid.) Yet, to make Gadamer's long story very short, the "baleful influence of language" is not immediately transparent. Language, as many note, reveals and conceals. The Great Divide conceals itself, resistant to criticism and transformation. While "averted vision" cannot itself stand outside language (there are no human positions outside language), such an indirect approach might help us begin to grasp the way in which *the Great Divide is schematized by language and perpetuated through the history of effects, even while the historicity and linguisticality of the schematization elude us*.[3]

My premise, then, is that averted vision allows us to see "the faint shadow" (like the fainter stars in the Pleiades) that the Great Divide itself casts on the body of nature, thereby disclosing its own existence. Once we see the shadow, and its source, we can then begin to bridge the Great Divide, so that the inner and outer are not antagonistic and hierarchical, but mutualistic and heterarchical. I begin with Darwin.

"Max," you may be wanting to say, "this is way too complicated. You've crossed over the line of academic convention (another boundary, eh?), which seeks clear and distinct ideas generated by an ironclad commitment to analytical method. You're relying way too much on metaphorical suggestion instead of conceptual analysis. And, frankly, you're violating 'the hidden codes' of human identity and dignity, which are all about rationality and control. So what's the need for this, what do call it, 'averted vision'? Talk about averting your gaze! I can't begin to fathom where you hope to go with this."

No doubt, you've got a case.[4]

Let me state a bold thesis, one that clearly sacrifices methodological rigor in favor of interpretive alternatives. The thesis has two parts. One is that the future of life itself, and with it the cumulative human achievements of several thousands, perhaps tens of thousands of years, are at risk. And the other is that even within those conversational circles accepting part one of my thesis, the outcomes are largely misdirected. Because of the Great Divide we do the wrong things. Not the right things.

This is not the occasion to reach for knockdown arguments. But consider that direct approaches to the resolution of a wide array of dysfunctional relations between cultural and natural systems, particularly our own industrial-consumerist society, are largely expensive failures. Mesmerized by the Great Divide, these actions focus on the obvious symptoms of environmental malaise, especially those visible at native-range human perception.[5] For example, tens of billions of dollars have been expended on environmental engineering and litigation under the provisions of CERCLA (the Superfund Law), while less than 100 million dollars have been spent annually by the U.S. government on pollution prevention. Or consider that several billion dollars annually are spent on fire suppression and exclusion on fire-adapted public lands forests, while only a few tens of millions per annum go toward restoration projects that would restore fire as a natural disturbance regime. Again, we treat symptoms, not the disease. More generally, proposed solutions for ecological dysfunctions are primarily driven by applied science and considerations of economic efficiency. Why, of course, would we expect any other strategy, given the Great Divide as it plays out in *the figuration of contemporary culture*? Culture and nature are bounded. Culture acts on nature, and not nature on culture. Nature is unruly, and must be cleaned up, brought under control, and otherwise rendered benign. Since we are separated from the natural world by the Great Divide, there appear to be no other alternatives.

Given the Great Divide, the multiple precipices of disaster—including the future of life, global climate change—that we presently skirt are no surprise. Thinking that nature is a goody box for exploitation, the human species has proceeded with aggressive abandon. Believing that nature is passively benign, and does not act back on the human estate, we conduct unprecedented experiments at unprecedented scales. And believing that the most fundamental meanings of life are compartmentalized within cultural cocoons, we dare not consider the idea that "the chthonian swamp" is our native home.

Yet we now know that the Great Divide is scientifically untenable and ethically bankrupt, a grievously mistaken account of the world, the things in the world, and appropriate relations between humans and the things in the world. Why then does the Great Divide continue to overdetermine the political process, economic policy, technological research, development and application, the curriculums of high schools and tier-one research universities, the allocations of funds from the NSF, and so on? Because we have failed to realize that this most elemental of boundaries has, through the history of effects, become a defining characteristic of the human condition.

More than 130 years ago, in the *Descent of Man*, Charles Darwin intuited that "the difference in mind between man and the higher animals, great as it is, is certainly one of degree and not of kind.... If it be maintained that certain powers, such as self-consciousness, abstraction, etc., are peculiar to man, it may well be that these are the incidental results of other highly-advanced intellectual faculties; and these are again *mainly the result of the continued use of a highly developed language*" [my emphasis] (1874, 106–107). Talk about the tip of the iceberg. Following the "genetic revolution," beginning with Mendel, then Watson and Crick, and continuing today in a variety of genome mapping projects, we are confronted with a stark truth: the human genome is virtually identical to that of the greater chimpanzee. There is less than one percent difference.

Darwin, despite his genius, cuts himself some slack, failing to grasp his own linguistic predicament. Ensconced inside scientific narrative, he turns antagonistic, and aggressively chides supernaturalists, which is to say those who clung to a boundary between God's chosen child, man, and the rest of the creation. What Darwin failed to see was that not only religion, and notions of Homo religiosus, but also science, and notions of Homo sapiens, were the artifacts of a "highly developed language." Had he followed through on his own linguistic turn, the outcome might have been different. He might have realized that all the various sciences are just as much human discourses, initiated through and sustained by language, as are the various religions.

The one percent of difference between human and chimpkind, as it turns out, makes all the difference. After all, chimpanzees are not cratering the planet. In effect, the specific difference between the human species and the other primates generally, and chimpanzees specifically, plays out in those capacities that make us language animals—Homo narrans. At the risk of dwelling on the obvious, some modest expansion is in order. Nobody knows it all. Everybody has a day job. So a Copernican-like revolution in human self-understanding that has taken place in the last fifty years is not necessarily common knowledge.

Following in Darwin's wake, much has happened. A few high points (low points?) are in order, beginning appropriately with evolutionary studies, such as those of Philip Lieberman (1984, 1998). He argues that a few biological changes explain the uniqueness of humankind. "Our present place in the scheme of things," he writes, "and the manifest differences between human beings and all other living animals regarding language and cognition follows from a sequence of small structural changes that together yield a unique pattern of linguistic and cognitive behavior" (1984, vii). However, Lieberman continues, "there is no linguistic 'gene,' nor is there a language 'organ' that can be localized in the human brain. The biological bases of human linguistic ability involve the characteristics of the central distributed cognitive neural computer as well as the species-specific neural mechanisms that are involved in the perception and production of speech and some aspects of syntax" (1984, 333). And finally, Lieberman argues, we cannot reduce the complexity of human behavior, such as ethical codes, to a biological basis. On the contrary, the complexities of culture and human behavior go far beyond any genetic explanation—an issue sometimes termed "the gene shortage problem."

Another dimension of the revolution in human self-understanding comes from archeological-anthropological studies, such as those of Richard Leakey (1992) and other "bone hunters." Leaky traces the natural history of Homo sapiens through the co-evolution of brain and language. One of the great mysteries of evolution has come to be called "explosive encephalization"—which I call the "monkey/human boundary," not altogether facetiously. Biological evolution typically works slowly, over hundreds upon hundreds of thousands of years. Yet over the course of perhaps 200,000-500,000 years, in the Homo habilis/Homo erectus time frame, there was a dramatic increase in cranial capacity, suggesting that the creatures lying along the trajectory of human descent were not only beginning to use tools but protolanguage as well. While there was no "revolutionary punctuation," Leakey argues, "the major event in the origin of modern humans very likely was the final acquisition of a fully articulate language" (1992, 274). There is virtually no doubt for Leakey that it was this perceptual/cognitive/linguistic evolutionary transformation that is the generative source for the cultural world in which we live and move and have our specifically human being. "Without language, the arbitrariness of human-imposed order would be impossible" (1992, 266). (While there is no direct evidence that brain size correlates with linguistic ability, the demands of neuromotor and information processing skills alone imply some relation.)

There are also a number of fascinating multidisciplinary, interdisciplinary studies that contribute to the revised picture of the human animal as Homo narrans. For example, the work of Terrence W. Deacon (1998), L. L. Cavalli-Sforza (1981; 2000), and Paul Ehrlich (2000). Deacon offers some conjectures especially relevant to grasping the Great Divide in *The Symbolic Species: The Co-evolution of Language and the Brain*. Part of his analysis conforms with "new paradigm" thinking. For example, he argues that "the ability to use virtual reference to build up elaborate internal models of possible futures, and to hold these complex visions in mind with the force of the mnemonic glue of symbolic inference and descriptive shorthands, gives us unprecedented capacity to generate independent adaptive behaviors" (434). However, this very capacity also underlies what Deacon calls "'symbolic compulsion.' "Ideologies, religions, and just good explanations or stories thus exert a sort of inferential compulsion on us that is hard to resist because of their mutually reinforcing deductive and inductive links. Our end directed behaviors are in this way often derived from such 'compulsions' as are implicit in the form that underlies the flow of inferences" (435). Arguably, the Great Divide plays out in culture as a symbolic compulsion. So construed, averted vision is linguistically based therapy hoping to disrupt the compulsion.

Paul Ehrlich's *Human Natures* is an evolutionary tour de force that is worth reading for a single conclusion alone. Namely, that evolutionary biology trumped itself with the human species, whose future for good or ill is tied to cultural evolution—deliberate or intentional cultural evolution. Among the many details of a rich and complicated argument, the so-called "gene shortage problem," Ehrlich argues, makes this so. "There are some 100,000 genes in the human genome, whereas there are roughly 100–1,000 trillion connections (synapses) between more than a trillion nerve cells in our brains [implying a ratio of 1 billion:1, synapses: genes].... Clearly, the characteristics of that neural network [mind↔brain] can only be partially specified by genetic information; the environment and cultural evolution *must* play a very large, often dominant role in establishing the complex neural networks that modulate human behavior" (124).

Cavalli-Sforza combines genetics, linguistics, and mathematical analysis in a remarkable way of thinking about cultural codes and their transformation. "From a genetic perspective," he writes, "our future is not terribly interesting—our species will probably not evolve much more" (2002, 205). The prospect for "endosomatic change" (i.e., change of the human genome) does not amount to much. But since the agricultural revolution, the rate of cultural change has been rapid. And, Cavalli-Sforza argues, "the rate of cultural change will continue to increase in the future" (206). But unlike genetic mutation, which is random and cannot be directed, cultural changes (evolution), Cavalli-Sforza contends (let us hope he is right) "are more often intentional or directed toward a very specific goal...while genetic change [is not]" (176).

Finally are anthropological-linguistic studies, such as those of Derek Bickerton (1990; 1995). Bickerton argues that humankind is "*not* unique in most of the ways in which we have been made out to be. That is, we do *not* come equipped with wisdom, logic, vast cognitive powers, novel problem-solving capacities, immaterial minds, immortal souls, and all the other baggage that has been imputed to us at one time or another. We have language grafted onto a primate brain, and that's it. We are still animals, but this two-edged gift that has been laid upon us obliges us to live in ways no other animal could conceive of" (1995, 156).

Following in Darwin's wake, I am claiming that no plausible understanding of humankind, either culturally or individually, can sunder itself from biological reality, including the anatomical and neurological bases of language. And, that said, no credible understanding of a cultural system—understanding by culture the accretion of information through language, that is, the culturgens or memes that direct human action within a cultural context—can therefore be sundered from biological underpinnings. Many have touched on this in a variety of ways. One is the philosopher, M. Merleau-Ponty (1968), who speaks of the awesome birth of vociferation, of the breath that, unlike the chirps and growls and bulges and sounds of all other creatures, becomes talk, of the talk that guides human action and becomes culture, of the culture that unintentionally becomes a plague upon the earth. But that is another story, elaborated in detail by many. More immediately relevant is the conjecture that the evolution, *ab initio*, of cultural forms generally and political-economic specifically is driven by demographics (see Johnson and Earle, 1987).

Jumping over fifteen thousand years of history, we come to the existing world order (which some have mistakenly characterized as the "New World Order": there's nothing new about something that has been thousands of years in the making). Try thinking of the existing world order—its legitimating narratives and political economy, technologies and industries, factories and agricultural fields, systems of communication, transportation, and finance, hospitals and schools, and so on—as the outcome of a long process of cultural evolution itself biologically driven by population growth. When there are mouths to feed, and more mouths on the way, humans are a clever lot. At least in the short term.

The Great Divide was *not intentionally hatched* in the minds of either prehistoric or historic ancestors. No grand narrative was constructed to guide humans in the exploitation of the earth's biophysical systems with neither concern for future generations of humans nor care for the rest of the creatures. Rather humans in particular places and at particular times were making decisions and taking actions that enabled material survival and offered psychological meaning. Cultural schemes that gave individuals a sense of purpose and that provided adequate material sustenance endured. Those that failed simply collapsed—either in the short run, through catastrophe (epidemics, famine), or in the long run, through the inability to adapt to changed circumstances, such as soil salinization due to irrigated agriculture.

The evolution of political economy, from band society to the urban-industrial state and the New World Order, has been biologically driven by the metabolic and psychosocial demands of ever-increasing numbers of humans. Cultural systems survived by evolving *ad seriatim* the political, technological, and other schemes, including codes governing sexual reproduction, necessary for environmental exploitation and maintenance of the social order. The present world order sustains six billion people, in varying degrees of material comfort, political freedom, psychological well-being, and physical health. If the objective was to create a world order that exploits the environment for the purpose of sustaining six billion humans for some indeterminate but clearly limited period of time, then we have succeeded. The biodiversity crisis was an unintended consequence. Human beings did not sit down around a collective table and intentionally decide to grow without limit, to convert the earth to one vast factory supporting themselves alone. But that has been the outcome.

However imperfectly, then, we can grasp the reality that the Great Divide itself casts a shadow on the body of nature.

To sum up, I have argued that language, codified into cultural narrative, mediates the uniquely human world, at one and the same time constraining human existence while also offering the possibility of freedom—the freedom to change the way humans interrelate with the living world, with each other, and indeed, self-identity itself. As a secondarily nidicolous species, we are born as wrinkled and tender little lumps of flesh, totally dependent on significant others to nurture and sustain us. Cultural transmission ensues even before birth (as with the preparation of a nursery in which the neonate will be socialized, the choosing of names, the planning for ceremonial introductions to the world, and so on), and continues throughout life. Along the way the growing child (and I speak primarily of the children of the West) is subtly conditioned into the webs of interlocution, including the narrative structures and themes that articulate the Great Divide.

A primary although not exclusive *source of human dignity*, of worth for Westerners, comes from a sense of control over destiny, over fate, over nature (see Taylor, 1985a; 1985b; 1989). Scientific narrative serves this end, as in the notion that through physics we will become the masters and possessors of nature, able to create a world to our liking. Psychological narrative serves this end, as with "the denial of death," which becomes a cultural causa sui that manifests itself in acquisitive materialism and commodity fetishism (see Becker, 1973). Yet another source of dignity is religious narrative, inherent in the notion that a people are chosen, or that time is a linear sequence going somewhere, or that there is a chosen, exceptional species. Political narratives also serve the same end, as in the Enlightenment texts that privilege the atomistic individual, and define a good society as one that economically satisfies the greatest number of individuals. Thus, I am suggesting that the Great Divide is cemented—scientifically, technologically, psychologically, economically, religiously—into our conceptions of self and society. We are caught, in narratives of our making. Nothing, it seems, can be changed. And yet transformation is possible.

There are a few "recursive thinkers" who move me again and again. One is Gregory Bateson (1972), who observes that "we are not outside the ecology for which we plan—we are always and inevitably a part of it" (504). Problem is, the Great Divide conceals that existential situatedness from us. Another recursive thinker, one who succinctly captures the predicament and the challenge inherent in bridging the Great Divide, is Marjorie Grene. She argues that a boundary—a.k.a. the Great Divide—created by and continuously reproduced through language, separates culture from nature. "We still have the image of a human world shorn of any roots in nature and a natural world devoid of places for humanity to show itself." The challenge, she argues (and I'm simplifying her argument), is to articulate how "historicity, as necessary condition for, and defining principle of, human being, can be within, not over against, nature" (1986, 182). In other words, if it is through language and its history of effects that we have been alienated from nature, then reconciliation might also be sought through language.

"All that you claim sounds sweet, Max," you might be thinking, "but it seems that you've succumbed to your Kansas childhood. Like Dorothy, you're dreaming. You think you've located the Emerald City, found Oz, and that the Wizard is going to save us all. But really, as you claim, it's just talk."

You may be right. Certainly about the Kansas roots. And, no doubt, the kind of analysis offered here is aptly characterized as "indeterminate scholarship," that is, arguments that cannot be tested empirically. Still, there is bigger context here: a legitimating narrative that begins with Darwin and the subsequent transformation of human self-understanding.

First, whatever the illusions of our culture (and every culture has its illusions), however beguiling the "Sirens of Scientific Power and Economic Progress, Forever," and regardless of the short-term (years, decades, centuries) successes of the controlling narratives, *ultimately those narratives are put to the test as a survival strategy*, not only for individuals, but ultimately at a cultural level, since decisions of individuals have aggregated, even synergetic, consequents. Just think of the so-called tragedy of the commons. Cavalli-Sforza's expositions are remarkably clear on the point. "Cultural selection can act counter to natural selection" (1981, 341). More recently he writes that "each cultural decision must pass through two levels of control: cultural selection acts first through choices made by individuals, followed by natural selection, which automatically evaluates these decisions based on their effects on our survival and reproduction" (2000, 178).

While Cavalli-Sforza ignores questions of temporal scale, the argument can easily be extended. By its very existence we can infer that any particular culture, such as our own, has been successful in the short run, allowing individuals to make rational short-term survival choices. This loaf of bread, this kind of automobile, this many children, and so on. However, the question of the century is cultural and natural sustainability, since individually rational choices in cultural context are not necessaryily sustainable in biophysical context. (For example, do any of us intend to voluntarily "get the price right" by paying a $5/gallon carbon tax next time we fill up with low lead? However ecologically sensible, such a self-imposed tax is economically irrational. Only a cultural intention can chance the calculus.)

And second, since language symbolically mediates the ongoing processes of cultural reproduction, *language also mediates the processes of cultural innovation*, including the theory and practice of sustainability. Adaptation to the exigencies of the ecosocial world in which we live and on which the future of life depends will either occur or not occur through the webs of interlocution that direct human behavior. Narratives are at one and the same time fate and freedom. The established stories constitute the cultural ground on which short-term survival depends, thereby limiting the ambit of immediate possibility. There can be no human reinvention *ex nihilo*, but only a future departing from where we are.

Yet the dominant stories can be teased apart and rewoven with new narrative themes among the old. We are, as has been said, a symbolic species. While we may not, because of symbolic compulsion (the mesmerizing, overwhelming psychosocial hold of exiting codes or co-adapted memes), we may also conceive, plan, and implement adaptive reconstructions. Derek Bickerton offers one of the clearest statements of this point.

> Language bestowed on... [humankind] powers that yielded far more than mere survival, powers that effectively conferred on our species the stewardship of the earth. Yet, formidable as these powers were, they carried within the seeds of destruction. Language has given us, not enough, but too much: not just the stewardship of the earth, but the capacity to destroy species weaker than ourselves, and even features of the environment on which our own survival might depend. Yet language is at the same time the nurturer and facilitator of all that is best in us, all that seeks to avoid such a fate and to bring us back into unity with the rest of creation. It is language, and language alone, that makes it possible for us to dream of a world of peace, freedom and justice where we might live in harmony with that nature of which, after all, we form only a dependent part. (1990, 257)

Clearly, you may disagree in principle with these two meta-moves. Language, as Kristeva reminds us, always remains partially unknown, for there is no specifically human position outside of language. But I ask with genuine (not Socratic) interest, if not these moves, then what are the alternatives? And please tell me, even provisionally, what capacities these alternatives offer for bridging the Great Divide?

Language, I have argued, is as real as the sounds we hear that bear human meaning, as strong as the phonemes and syntax that carry it, and as ephemeral as the small muscle movements and neural nets that direct it. Altogether it constitutes the medium in which and through which we live. Through language and only through language might we loosen the hold of the Great Divide and find our way into a future that is for the moment powerless to be born.

Notes

1. See Donna Haraway (1997, 11). "Figurations are performative images that can be inhabited. Verbal or visual, figurations can be condensed maps of contestable worlds. All language, including mathematics, is figurative, that is, made of tropes, constituted by bumps that make us swerve from literal-mindedness. I emphasize figuration to make explicit and inescapable the tropic quality of all material semiotic processes.... For example, think of a small set of objects into which lives and worlds are built—chip, gene, seed, fetus, database, bomb, race, brain, ecosystem. This mantralike list is made up of imploded atoms or dense nodes that explode into entire worlds of practice. The chip, seed, or gene [and ecosystem] is simultaneously literal and figurative. We inhabit and are inhabited by such figures that *map universes of knowledge, practice, and power*" [my emphasis].

2. The underlying premise can be characterized in different ways. Kristeva (1989), for example, argues that language always remains unknown because we can

never get outside of language. To take another example, Heidegger argues that language always and simultaneously reveals and conceals. And perhaps best of all is the ancient aphorism that just as the fish is the last to discover it swims in water, so, too, humankind and language.

3. More directly stated, "reflexive questions," Hilary Lawson writes, "have been given their special force in consequence of the recognition of the central role played by language, theory, sign, and text. Our concepts are no longer regarded as transparent—either in reflecting the world or conveying ideas. As a result all our claims about language and the world—and implicitly all our claims in general—are reflexive in a manner which cannot be avoided. For to recognize the importance of language is to do so within language" (1985, 9).

4. I've been reading Paul Feyerabend (1989) for too many years. And more recently Richard Bernstein (1983).

5. Neil Evernden's (1985) insight into the conceptual trap of environmentalism remains unexceeded. "We call people environmentalists because what they are finally moved to defend is what we call environment. But, at bottom, their action is a defense of cosmos, not scenery. Ironically, the very entity they defend—environment—is itself an offspring of the nihilistic behemoth they challenge. It is a manifestation of the way we view the world" (124).

References

Bateson, Gregory. 1972. *Steps to an Ecology of Mind*. New York: Ballantine.

Becker, Ernest. 1973. *The Denial of Death*. New York: Free Press.

Bernstein, Richard J. 1983. *Beyond Objectivism and Relativism: Science, Hermeneutics, and Praxis*. Philadelphia: University of Pennsylvania Press.

Bickerton, Derek. 1995. *Language and Human Behavior*. Seattle: University of Washington Press.

———. 1990. *Language and Species*. Chicago: University of Chicago Press.

Cavalli-Sforza, L. L., 2000. *Genes, Peoples, and Language*. New York: North Point Press.

Cavalli-Sforza, L. L., and M. W. Feldman. 1981. *Cultural Transmission and Evolution: A Quantitative Approach*. Princeton: Princeton University Press.

Darwin, Charles. 1874 (1871). *The Descent of Man, and Selection in Relation to Sex*. London: John Murray.

Deacon, Terrence W. 1998. *The Symbolic Species: The Co-Evolution of Language and the Brain*. New York: W.W. Norton.

———. 1995. *Language and Human Behavior*. Seattle: University of Washington Press.

Ehrlich, Paul. 2000. *Human Natures: Genes, Cultures, and the Human Prospect*. Washington, DC: Island Press.

Evernden, Neil. 1985. *The Natural Alien: Humankind and Environment*. Toronto: University of Toronto Press.

Feyerabend, Paul. 1989. *Against Method*, rev. ed. New York: Verso.

Gadamer, Hans-Georg. 1988. *Truth and Method*. Trans. Garrett Barden and John Cumming. New York: Crossroad Publishing.

Grene, Marjorie. 1986. "The Paradoxes of Historicity." In Brice R. Wachterhauser, ed. *Hermeneutics and Modern Philosophy*. Albany: State University of New York Press.

Haraway, Donna J. 1997. *Modest_Witness@Second_Millennium. FemaleMan ©_Meets_OncoMouse™: Feminism and Technoscience*. New York: Routledge.

Johnson, Allen, and Timothy Earle. 1987. *The Evolution of Human Societies: From Foraging Group to Agrarian State*. Stanford: Stanford University Press.

Kristeva, Julia. 1989 (1981). *Language, The Unknown: An Invitation into Linguistics*. Trans. Anne M. Menke. New York: Columbia University Press.

Lawson, Hilary. 1985. *Reflexivity: The Postmodern Predicament*. LaSalle, IL: Open Court.

Leakey, Richard, and Richard Lewin. 1992. *Origins Reconsidered: In Search of What Makes Us Human*. New York: Doubleday.

Lieberman, Philip. 1984. *The Biology and Evolution of Language*. Cambridge: Harvard University Press.

———. 1998. *Eve Spoke: Human Language and Human Evolution*. New York: W.W. Norton.

Merleau-Ponty, Maurice. 1968. *The Visible and the Invisible*. Trans. Alphonso Lingis. Claude Lefort, ed. Evanston, IL: Northwestern University Press.

Petersen, Anna L. 2001. *Being Human: Ethics, Environment, and Our Place in the World*. Berkeley: University of California Press.

Taylor, Charles. 1985a. *Human Agency and Language: Philosophical Papers 1*. Cambridge: Cambridge University Press.

———. 1985b. *Philosophy and Human Sciences: Philosophical Papers 2*. Cambridge: Cambridge University Press.

———. 1989. *Sources of the Self*. Cambridge: Harvard University Press.

Chapter 2

Lamarck Redux: Temporal Scale as the Key to the Boundary Between the Human and Natural Worlds

J. Baird Callicott

The Boundary Idea at the Dawn of Western Philosophy

Arguably the most ancient concern in Western philosophy is boundaries and the lack thereof. Inferring from his answer, Thales—according to Aristotle, the first person to concern himself with questions about nature instead of the gods—posed the first (implicit) question of Western philosophy: Of what is the world composed? His successor, Anaximander, rejected Thales's answer, Water, in favor of what he called the ἄπειρον. This word is formed from πέρασ, meaning *end, limit,* or *boundary,* in combination with the alpha privative (α-) meaning *-less, un-* or *non-.* The cosmic ἀρχή—that out of which everything else is composed, the stuff of the world—is, according to Anaximander at the dawn of Western philosophy, the ἄπειρον, the endless, the unlimited, the unbounded. This word, pregnant with so much significance, is often anachronistically mistranslated as the "infinite." The idea of a spatially infinite substance could only arise a century later in dialectical response to conundrums proffered by Parmenides and Zeno. Though we can never know for sure, Anaximander probably meant to suggest that unlike a common substance—such as water, which is wet, or air, which is dry—the ἀρχή can have no definite qualities and was homogeneous. The ἀρχή must be ἄπειρον, that is, an indefinite stuff without internal boundaries.

Being alive and divine, Anaximander's ἄπειρον began to move itself in a swirl or vortex motion, and thereafter more definite stuff—the hot, the cold, the wet, and the dry—"separated out" of it. These more definite substances—

later reified by Empedocles as fire, earth, water, and air, respectively—were gathered, like to like, into regions. Earth collected in the center of the swirl, surrounded by a hydrosphere, an atmosphere, and a pyrosphere. Naively experiencing our world, unprejudiced by post-Copernican astronomy and geography, we do indeed observe the primary substances of the world to be internally bounded, albeit porously and imperfectly, in just such a way. Thus, at the inception of systematic Western thought about nature, it seems that what was believed to be essential to the formation of a κόσμοσ—a beautiful world order—is the establishment of πέρατοι, boundaries. Indeed, the good left-hand column of the Pythagorean Table of Opposites begins with πέραο opposed, in the bad right-hand column, by ἄπειρον. The Limit and the Unlimited, the Boundaried and the Unboundaried are, for the most ancient of Western philosophers, the first, the most primitive of cosmic principles.[1]

Perhaps as a legacy of this most venerable tradition of Western philosophical thought, the concept of a boundary is primarily spatial, even more particularly geographical. As Western philosophy progressed from primarily material concepts (concern with the "material cause" as Aristotle catalogued it) to more abstract concepts (concern with the "formal" cause as Aristotle catalogued it) boundaries between things were thought to be drawn by their forms or essences. Thus, what separates one species from all the others was its essence, according to both Plato and Aristotle. Famously, what distinguishes the human species from all the others is reason: "man" (ἀνθρωποσ) is the rational animal. Reason is man's essence.

Perhaps not surprisingly, the spatial connotation of "boundary" is primary (because primal?) in the Western mind. The combination of spatial and essential connotations leads to a derivative political connotation. Thus, when we hear "boundary" we are liable first to think spatially of frontiers, borders—often defined geographically by rivers, mountain ranges, sea coasts—that separate nation-states from one another. The nation-state is the territory (the bounded geographical space) of a people separated from other peoples by a set of essential characters (race, a common language, history, culture). Crossing or penetrating boundaries strongly suggests, though it certainly does not actually entail, locomotion, that is, motion from one spatial/geographical location to another.

One objective of the ideology of shifting political identity in a multicultural, pluralistic nation-state, such as the United States, is to define the boundary (the essence) of the group— American Indian, African American, Hispanic, Latino/a, Chicano/a, Gay, Lesbian—who is in and who is out. Sometimes hardening the boundaries of identity politics threatens to fragment nation-states into quasi- or fully independent territories, even new states. The Basque independence movement in Spain and the Kurdish independence movement in Turkey, Iraq, and Iran are examples; and today we hear much about the possibility of a new Palestinian state. In such cases, the metaphorical boundaries of identity are translated into paradigmatic spatial/geographical boundaries.

Thus, there remains, at least as I register it, a spatial/geographical/essential residue, in political connotations of the boundary idea. The boundaries of identity politics *in extremis* become lines drawn on a map and fences and checkpoints and border crossings on the ground. But even when political identity does not run to such extremes, essential boundaries are translated into paradigmatic spatial boundaries. In Beirut, a "green line" separates Muslim from Christian quarters of the city. In every metropolis in the United States there is the ghetto, the barrio, the hood, the gay district. In the 1960s Chicano/a essentialists envisioned a separatist homeland in the American southwest, named Aztlan. The spatial boundaries between states may shift with the meandering river separating them or with the de facto loco-movement of peoples from one territory of a state to that of another, as Albanians eventually displaced Serbs in Kosovo and Mexicans are displacing Anglos on the U.S. side of the Rio Grande in Texas. Essential boundaries separating identity groups, as gays and lesbians, for example, unite with one another and with bi- and transsexuals.

Here I depart from the deeply ingrained spatiality of the boundary idea in Western thought—with all its permutations and metaphorical extensions—and explore ways in which temporal scale creates boundaries that we may not be prepared to recognize as such (precisely because of the deep spatial bias in our boundary thinking). Sometimes, indeed, temporal scale creates spatial boundaries, but often it creates essential boundaries. Rather , I shall argue, more precisely, that at least in the case of "man" (ἄνθρωποσ) the alleged essential boundary is really a temporal boundary. My ultimate goal is to resolve a contemporary conundrum of environmental philosophy. Are we human beings set apart from nature by some kind of essential boundary? Or are all such putative boundaries between people and nature only obsolete theological and philosophical fictions?—in which case man is a part of nature. I argue that Homo sapiens is a natural species, but also that while Homo sapiens and human culture evolved in a Darwinian manner, the emergence of culture as a biological adaptation has caused human evolution to take a quantum leap, to use an overworked metaphor, into a novel Lamarckian evolutionary orbit. And I argue that the temporal scale of the Lamarckian evolutionary ambit in which Homo sapiens now adapts to and transforms the natural environment creates a boundary between our species and all the others—that is, a boundary between us human beings and the rest of nature. I start with a contemporary classic discussion in ecology of how spatial boundaries emerge at the interface of temporal scales.

How Spatial Boundaries Emerge at the Interface of Temporal Scales

In a deservedly famous paper, "Cross-scale Morphology, Geometry, and Dynamics of Ecosystems," C. S. Holling sets out to explain a curious phenomenon of nature. The body masses of terrestrial birds and mammals, irrespective of cardinal guild (i.e., whether the animals are herbivores, omnivores, or carnivores), are "clumped." In other words, there are gaps in the distribution of their sizes. Or, put the other way around, their body masses cluster around certain values. In the plainest of terms, the sizes of birds and mammals are not distributed evenly from small to large; rather, they come in eight basic sizes—from very small (hummingbirds and chickadees) to small (wrens and sparrows) to several classes of medium (robins, blue jays, crows) to large (hawks and owls) to very large (egrets, herons, cranes, flamingos). Thus, we say that there are boundaries between the sizes of birds and mammals. Holling argues that these boundaries exist because animals are adapted to discontinuous small, medium, and large home-range ecosystems that are hierarchically ordered—that is, the small-scale ecosystems are nested into the medium and the medium into the large-scale. Animals adapted to the inner, small-scale ecosystems, cluster around small-size values and those adapted to the large-scale ecosystems cluster around large-size values.

Ecosystems, however, are not constituted by an interacting suite of organisms, but by a linked suite of processes. Organisms, rather, carry out some ecosystem processes, such as photosynthesis and nutrient cycling. Because they are composed of processes, not organisms, the boundaries of ecosystems are determined by the frequency or temporal scale of the processes that compose them. As Holling puts it, "The landscape is hierarchically structured by a small number of structuring processes into a number of nested levels, each of which has its own physical textures and temporal frequencies. That is, the processes that generate discontinuous time dynamics also generate discontinuous physical structure."[2] Thus, the temporal scale of ecological processes determines the spatial scale of habitats and the boundaries between them, which in turn determines the boundaries between the body sizes of birds and mammals.

According to Holling, the way animals perceive their environments plays a critical role in the evolutionary selection of their size category: "A hierarchical organization in a landscape generates abrupt shifts in the kind, distances, and size of objects when the grain of measurement reaches a value that aggregates objects into a new set of objects in a new hierarchical level. Since the size of the animal defines the grain of its measurement, then a body-mass gap exists at those sizes and sampling grains where there is an abrupt transition in the attributes of objects (their size and inter-object distance) between two hierarchical levels."[3] By way of illustration—mine, not Holling's—a mouse may be able to see individual trees, but not a patch of woods; whereas a squirrel can see not only individual trees, but patches of woods, and determine whether the distances between woody patches is close enough to attempt safely to run between them. A coyote can not only see but mentally map a congeries of landscapes, each of which is composed of multiple patches. The discontinuity between the sizes of "objects" at these several hierarchical levels—trees at one level, woody, grassy, and shrubby patches at the next level up, and the landscape composed of many patches of different kinds at the next level up from that—is mirrored by the discontinuity of the sizes of animals.

Identifying Temporal Scales of Interest to Ecological Philosophy

We live in a world riddled with processes going on at various temporal scales some of which structure our lives in important ways and some of which do not. The diurnal cycle—the alteration of night and day—structures almost everyone's life in extremely important ways. The four-year cycle of the Olympic games structures the lives of Olympic athletes and trainers, and the planners, vendors, and so on who are professionally involved with the Games, while, for the rest of us, that cycle is of little structural importance for our lives and of only passing interest. Some of the important ways processes going on at various temporal scales structure our lives are interesting for some purposes, but not for others. The roughly ten-year business cycle of the U.S. national economy—growth, punctuated by stagnation and recession—more or less structures the economic lives of every American and is thus of interest to contemporary American environmental philosophers, not, however, qua environmental philosopher, but qua consumer and future retiree. Finally, some processes going on at various temporal scales may not structure anyone's life in any appreciable way, but may be of interest nevertheless. The period of the orbit of Mars in relation to the orbit of the Earth may not structure anyone's life, but for centuries it has been of intense interest to astronomers.

Of interest to ecological philosophy, "At least six hierarchical levels can be identified, each of which is dominated by one category of structuring processes," according to Holling. Curiously, however, Holling goes on to identify only three: "The smaller and faster scales are dominated by vegetative processes, the intermediate by disturbance and environmental processes, and the largest and slowest by geomorphological and evolutionary processes."[4] I will try to elaborate and refine Holling's insight.

The Vegetative (Organismic) Temporal Scale

What are vegetative processes and what sort of temporal scales do they define? Obviously, first, there is photosynthesis, which occurs in the daylight hours and ceases at night. Also in temperate latitudes photosynthesis is seasonally pulsed. Thus, there are growing seasons of varying length, followed by correlative dormant periods, on an annual cycle. Dead leaves and other detritus fall to the ground and are there decomposed by fungi, worms, and bacteria—also roughly on an annual cycle in temperate latitudes. If the patch is a woods, another important process is crown formation, which, depending on species and other factors, may occur in one decade or several. This process scales up from annual vegetative processes by an order of magnitude. At the next order of magnitude, over about a century, depending on species, trees reach their maximum height, and may live for several more centuries.[5]

So the scale range of vegetative processes runs from the day-long to the season-long to the year-long to the decade-long to the century-long up to and rarely exceeding the millennium-long. Holling calls this temporal scale "vegetative," I surmise, because plants dominate the landscape and, as autotrophs, provide the basis of almost all Earth's trophic pyramids. More generally, we might call this the "organic" temporal scale and recognize our own lifetimes—now often lasting eight or nine decades—as spanning a significant portion of it.

The Ecological Temporal Scale

The next temporal scale range that Holling identifies he leaves unnamed, but I suggest we call it the ecological temporal scale. One fundamental ecological process is succession. Suppose a forest patch is clear-cut or windthrown and left to recover on its own. The first to colonize the cleared ground would be sun-loving herbaceous annuals, followed by weedy perennials. After a decade or so these would be overtopped and replaced by short-lived brush, shrubs, and scrubby trees. After several decades the brush, shrubs, and scrubs would be replaced by longer-lived, taller trees. And over a period of centuries the forest would gradually change in composition as shade-tolerant tree species replaced species whose seeds cannot germinate under closed canopies. So, all told, the temporal scale of ecological succession runs from decades to several millennia.

Disturbance regimes also constitute an important set of ecological processes. In some ecosystems fires occur periodically: every decade or so, at the greatest frequency; every twenty to fifty years; every 200 years; it varies. Most riparian ecosystems are subject to flooding, often annually, but in some riparian ecosystems there are floods of greater magnitude that occur less frequently: once every fifty years, on average; once every hundred. Coastal plains are periodically disturbed by hurricanes, which come ashore several times a year to the Atlantic and Gulf coasts of North America, but to the same swath only every hundred years or so, which is often enough to affect the composition of coastal biotic communities and the structure of coastal ecosystems. The North American plains are also subject to violent disturbance by wind. Another periodic ecological process is population cycles in animals, which usually occur in periods not less than a decade nor more than several decades in length.

From these examples of important ecological processes—succession, disturbance regimes, population cycles—ecological temporal scales appear to range from periods of a decade to several thousand years.[6]

The Evolutionary Temporal Scale

It seems to me that Holling's crude category of the "geomorphological and evolutionary" temporal scale should be divided in two and reversed in order as we move from consideration of smaller to larger temporal scales. How long does it take for species to evolve by natural selection? What is the natural life span—from origin to extinction—of a naturally selected species? The answer to these questions depends on the species. Some insects can segregate into races and evolve into full species over a few hundred generations in only ten years. A diatom endemic to Yellowstone Lake evolved from its parent (now sister) species over several centuries and has remained stable for 8,000 years. The charismatic megafauna on which so much conservation effort is lavished required hundreds of thousands of years to evolve into what they presently are. We Homo sapiens have remained anatomically and physiologically stable for somewhere between 100,000 and 200,000 years. The genus Homo has existed for somewhere between 1.5 and 2.5 million years. The 1,100 or so extant species of sharks have existed on average for about 100 million years. The average life span of a vascular plant species or vertebrate animal species is about a million years. So the evolutionary temporal scale (ignoring outliers such as rapidly evolving insects and the incredibly long-lived family of sharks) appears to range from 100,000 to several million years.[7]

The Geomorphological Temporal Scale

The temporal scale of geomorphology—the processes shaping the face of the Earth—is much greater still. By far the most potent earth-shaping process is plate tectonics, which is a manifestation of an even more fundamental geomorphological process, the rock cycle. The Earth's lithosphere is broken into about fifteen plates that float on the denser molten material beneath them and move, relative to one another, in different directions. When these moving plates collide they cause continental plains to rise in elevation and, more extremely, mountain ranges to thrust up. For example, the eastward moving Nazca plate colliding with the westward moving South American plate created the Andes; the Himalayas were formed by the northeastward moving Indian-Australian plate colliding with the southwestward moving Eurasian plate. The plates move at rates of between 2 to 15 centimeters per year. At this pace, it takes about fifteen million years to build a mountain range, such as the Sierra Nevada. At the boundary between colliding plates, such as the northwestward moving Pacific plate and the southeastward moving Eurasian plate, the heavier, lower-elevation rock of the oceanic plate is forced back down into the molten rock below, as the continental plate is lifted up.

Molten igneous rock oozing from ridges in the middle of the Atlantic, the eastern Pacific, the Indian and Southern oceans cooling and solidifying, forces the plates on either side to move in opposite directions. The elevated continental rocks weather—from the action of wind, water, and carbonic acid—creating sediments that gravity slowly carries into the oceans. It takes about 100 million years for a mountain range to erode away. Adding to the weathered sediments are calcium carbonate biosediments produced by shellfish and corals. These sediments accumulate and are compacted by more sediments washing in on top of them, eventually forming limestone and other sedimentary rocks. As the process of deposition continues, pressure and heat transform sedimentary into metamorphic rocks. This closely coupled set of process—spreading, plate movement, subduction, uplift, weathering, and sedimentation—constitutes a cycle in which the Earth's rocky mantle actually turns over completely. But the process is so slow that the Earth, in its 4.5 billion year existence, has only completed one full rock cycle. Thus, the geomorphological temporal scale appears to range from hundreds of thousands to millions up to a couple billion years.[8]

The boundary at the interface of the geomorphological temporal scale of plate tectonics and the vegetative (or more generally organismic) temporal scale of human affairs is so profound that one pair of writers make it a joking matter: "The Pacific plate is moving north relative to the North American plate at a rate of approximately 5 cm/year.... As a result, Los Angeles, now more than 500 km south of San Francisco is moving slowly toward that city. If this motion continues, in about 10 million years San Francisco will be a suburb of Los Angeles."[9] Of course in ten million years, Homo sapiens will almost certainly be extinct and all traces of San Francisco and Los Angeles will have weathered away and washed into the eastern Pacific sediment basin. On the other hand, the boundary at the interface of the geomorphological and the organismic temporal scales is by no means impenetrable. For the inching northward of the Eastern Pacific plate against the North

American along the San Andreas fault (the spatial boundary between the two) causes numerous earthquakes and occasional volcanic eruptions (such as the recent Mount Saint Helens eruption in Washington) at frequencies that fall well within the range of the organismic temporal scale.

The Climatic Temporal Scale

Although not mentioned by Holling, the temporal scale of climate change should also be of particular interest to ecological philosophy. Especially important to the proliferation of biodiversity, which peaked in the Pleistocene, has been the cycle of glaciation and glacial interstadials during that geologic epoc—because a significant factor in speciation is geographic isolation. The isolation of continents and large islands produced by plate tectonics, for example, allowed very different fauna to evolve in the Americas, Australia, Eurasia, and Africa—a case of interpenetration at the boundary of the geomorphological and evolutionary temporal scales. Glaciation driven by climate change has produced similar isolation as single species are driven into small unconnected refugia, by either the advance of glaciers or the associated desiccation of the climate, there to evolve independently into different species. In this instance, we have an interpenetration at the boundary between climatic and evolutionary temporal scales.

The major factor influencing global climate change appears to be changes in Earth's average temperature, which are notoriously unpredictable and may occur rather suddenly. For example, a sudden drop in temperature lasting 300 years—1550–1850 CE—had such an impact on the climate that it is sometimes called the Little Ice Age. It was too brief to have any notable evolutionary significance, but it may have lasted long enough to have had some ecological significance. If so, in this instance, we have an interpenetration at the boundary between the climatic and ecological temporal scales.

What the Earth's climate was in the distant past is confounded by many uncertainties. It seems that during most of Earth's biography, global temperatures were probably warmer than they are presently. During the last billion years, it appears that glacial pulses began at roughly 925, 800, 680, 450, 330, and 2 million years BP. Of these pulsations, the most severe seems to be the one that began 800 million years ago, during which glaciers may have come within 5 degrees of the equator. The most recent glacial pulsation, the Pleistocene, began two to three million years ago and is characterized by four major long periods of glaciation—reaching 40 degrees latitude—punctuated by shorter interstadials. It began to end, if indeed it is ending, 14,000–16,000 years ago.[10]

So what are the parameters of the climatic temporal scale? Getting a grip on this temporal scale seems more complicated than getting one on the organismic, ecological, evolutionary, and geomorphological. For the concept of climate includes not only temperature, but rainfall and seasonality.

At its upper end, defined by slower and spatially global climate change—such as glacial advance and retreat—climatic temporal scale is easier to grasp. The Earth as a whole has no seasons and—assuming average constant elevation of land forms at the geomorphological temporal scale—global rainfall appears to be a function of global temperature: the colder the global climate, the drier. For as more of the Earth's water is sequestered in glacial ice, less is evaporated from the diminished oceans, and the colder atmosphere holds less moisture. So when it gets cold enough for glaciers to advance, the global climate also gets drier. At the lower end of the climatic temporal scale and at regional spatial scales rainfall and seasonality greatly complicate the picture. The difference in the climate, for example, of Louisiana and West Texas is more a matter of rainfall than temperature. And the difference in the climate of Dallas, Texas. and Seattle, Washington. is more a matter of seasonality, for both receive about the same amount of rainfall.

Reflecting on its deep grammar, "climate" is not subject to annual fluctuation. For example, we might well say of the recent New England climate that it is characterized by long, cold winters, but not that it radically changes from winter to summer. Average global and regional temperatures do fluctuate slightly from year to year; and they do so noticeably from decade to decade. So does average regional rainfall. Still, we would not say that the New England climate has changed if several years of severe winters are followed by several years of mild winters or if several wet years are followed by several dry. But during the 300 years of the Little Ice Age, we would say that a change in the New England climate then occurred. So it seems that the bottom range of the climatic temporal scale is measured in centuries. The cycle of the glacial incursion and retreat during the Pleistocene occurred over thousands of years and may still be going on—that is, we may at present be living in an interstadial to be followed in the future by another more prolonged Ice Age. The cycle of Pleistocene-scale glacial pulsations occurred over millions of years. This suggests that the scale of global climate change is very broad and is itself hierarchical. At the base of the hierarchy are changes that register only after a century or more; still more severe fluctuations register after thousands of years; and so on, through several more orders of magnitude.

How Processes at Different Temporal Scales Interact at the Boundary

The world is dynamic at every temporal scale, but the processes occurring at slower temporal scales can be regarded as stable (unchanging) at the boundary or interface with processes occurring at more rapid temporal scales. For example, if an ecologist is studying the interacting population dynamics of the snowshoe hare and lynx, both of which cycle, he or she may regard the altitude and latitude of his or her study areas to be constant. Both are, in fact, changing. The altitude of Canada, where most lynx and snowshoe hare reside, is rising as the Earth's mantle rebounds from the weight of the glacial ice that burdened northern North America 14,000 years ago. But that climatic-geomorphological process is occurring at such a slow temporal scale that changes in elevation measured at the ecological temporal scale of population dynamics and the organismic temporal scale of the ecologist's research grant will be so slight as to be an insignificant factor for lynx–snowshoe hare population dynamics. Similarly, the North American plate is moving and so the latitude of Canada is changing, but again, relative to ecological and organismic temporal scales, so slowly as to be of no more significance to ecology than the future proximity of Los Angeles and San Francisco is to the real estate industry.

Processes at slower temporal scales also constrain processes at faster temporal scales. For example, the relatively constant climate of a region constrains processes going on at the ecological scale in that region. In a region with, say, a four season (spring, summer, fall, winter) temperate climate, the potential species diversity of biotic communities is more limited than in a region with a two season (wet and dry) tropical climate. And, for another example, processes occurring at ecological temporal scales constrain processes occurring at organismic temporal scales. In a locale with, say, a frequent fire regime (which is, in turn, climatically constrained) trees may be prevented from reaching maturity and reproducing.

Processes occurring at faster temporal scales are damped out before they reach distant boundaries; they do not cross remote borders. For example, the most industrious human effort to locally influence geomorphology—dam building, strip-mining, stream straightening—will have no measurable effect on plate tectonics or the rock cycle. Processes occurring at faster temporal scales are sometimes essential to the function and persistence of processes occurring at proximate slower temporal scales. Obviously, for example, plant growth and reproduction, occurring at the organismic scale, are essential to succession and, as fuel, to fire regimes occurring at the proximate ecological temporal scale. But processes occurring at slower temporal scales are normally unresponsive to the *vagaries* of processes occurring at more rapid temporal scales. The often vital functional effects of processes occurring at faster temporal scales on those at proximate slower temporal scales are modulated through averaging as they migrate across the border. For example, the life cycle of individual trees in a temperate forest occurs at the higher end of the organismic temporal scale, and at a relatively small, gap-sized spatial scale of say a maximum radius of 15 meters. The life cycle of individual trees is registered at ecological temporal and relatively large landscape spatial scales as an average: so many dead and down trees, so many new seedlings of some species sprouting under the canopy and of other species in the gaps per hectare per year. In general, at ecological temporal scales, change in the overall composition and structure of the forest is driven in part by processes occurring at organismic scales, but the stochasticity of the events at the organismic temporal scale (more dead and down trees this year, fewer next; more dead and down trees in this hectare, fewer in that) is averaged out at the ecological temporal scale.

Sudden changes in that rate of processes occurring at one scale can storm across the border like an invading army and appreciably alter processes at neighboring scales.

Ecologists usually assume that the climate of their study areas is changing so gradually as to be—relative to the ecological temporal scale—as constant or stable as latitude and elevation. Indeed, F. E. Clements, the dean of early twentieth-century ecology, assumed that regional climate was unchanging, relative to the ecological temporal scale, and thought that stable regional climate was the sole determinant of the climax community—the putative end point of ecological succession. (He chose the name "climax" for this hypothetical point of successional equilibrium precisely to link it with "climate" semantically.) However, if Clements were investigating ecological succession in the middle of the Little Ice Age, then the sudden cooling might have appreciably altered community composition and succession, for that three hundred year period, especially at the spatial boundaries between biotic communities called ecotones. That is an example of a rate change so sudden at a slower level in the hierarchy of temporal scales that it impinges on processes going at the next faster level. A temporal border is invaded.

Conversion of the Amazonian rain forest to brush and pasture on a large spatial scale—an anthropogenic ecological process; succession in reverse, as it were—can affect the otherwise slow-changing regional climate. For much of the atmospheric moisture (and thus rainfall) in the Amazon basin is produced by transpiration. Reduced forest cover→reduced transpiration→reduced atmospheric moisture→reduced rainfall = regional climate change. Before the advent of industrial logging and cattle ranching, a little clearing here, a little more there, a little more this year, a little less next year, gets averaged out and damped down in the Amazon when it registers at the temporal scale of regional climate change. Presently, the reverse successional processes occurring at the ecological temporal scale in the Amazon basin has become so accelerated and spatially widespread that they cease to be damped down and averaged out—and thus threaten to initiate a change at the next higher temporal scale, that is, to initiate a change in the regional climate.[11]

The Erased Boundary Between the Human and Natural Worlds in Aldo Leopold's Environmental Philosophy

Controversy and confusion about the boundary between the human and natural worlds confound many domains of ecological philosophy from environmental ethics to conservation theory. Aldo Leopold, for example, seems to locate the human in the natural world, that is, to erase what may have seemed to him to be an indefensible essential boundary between the two drawn in traditional Western religious and philosophical ideology. Taking the anthropological consequences of the theory of evolution seriously, Homo is but one extant genus of anthropoid ape among four others—a monkey's uncle (our cousin), so to speak. As Leopold put it, "it is a century now since Darwin gave us the first glimpse of the origin of species. We know now what was unknown to all the preceding caravan of generations: that men are only fellow-voyagers with other creatures in the odyssey of evolution. This new knowledge should have given us, by this time, a sense of kinship with fellow creatures."[12]

This evolutionary naturalization of the human sphere is a cornerstone of Leopold's land ethic, which "changes the role of *Homo sapiens* from conqueror of the land community to plain member and citizen of it."[13] For if Homo sapiens were not, in fact, a plain member and citizen of the biotic community then he would, Leopold seems to have supposed, be under no obligation to evince "respect for...fellow-members, and also respect for the community as such." That is because Leopold also borrowed from Darwin the idea that "all ethics so far evolved rest upon a single premise: that the individual is a member of a community of interdependent parts."[14] If this is true, then, if one is not a member of a given community (biotic or otherwise), one has no duties to fellow members of that community (qua that community) or to that community as such.

Peter Fritzell points out the contradiction in Leopold's argument: "To be a part, yet to be apart; to be a part of the land community, yet to *view* or *see* oneself as a part of that community (and thus to remain apart from it)— that is the dilemma.... If man is a plain member and citizen of the land community, one of thousands of accretions to the pyramid of life, then he cannot be a nonmember and conqueror of it; and his actions (like the actions of other organisms) cannot but express...his position within the pyramid of life."[15] That is, if people are plain members and citizens of their biotic communities, then what people do in and to their biotic communities as such and to their fellow members of those communities is amoral—no different in this regard from the destruction of African forests by elephants or of American forests by gypsy moths. Conversely, if people are not plain members and citizens of proximate biotic communities, then they have no obligations to their nonfellow nonmembers and to their noncommunities as such. So have it either way—Homo sapiens either is or is not a part of nature—an environmental ethic of the kind Leopold envisioned is conceptually incoherent.

Fritzell's conception of the boundary between the human and natural worlds, on reflection, seems quaintly classical—harking all the way back to the tree of the knowledge of good and evil in the Bible. Upon eating its fruit Adam and Eve could *view* or *see* themselves and discover that they were naked. This unique human capacity for selfconsciousness is the fountainhead, Fritzell seems to suggest, of both disinterested science and compassionate ethics.

Thus, as Fritzell unwittingly reminds us, the boundary between the human and natural worlds was classically drawn by appeal to some *essential* difference. Human beings are uniquely created in the image of God, or are uniquely self-conscious, or uniquely rational. That tradition has survived into the contemporary debate about the existence and location of that Great Divide when the putatively unique human capacities for language use and/or tool making are enshrined as boundary markers (or makers).

Also, without a sharp boundary between the human and natural worlds, conservation biology lacks a clear notion of what to conserve. Biological diversity, the most ubiquitously evoked target of conservation efforts, can in principle to enhanced anthropogenically by introducing species to communities with empty niches—which would be anathema to most conservation theorists. And from a post-Darwin Leopoldian perspective (the view that Homo sapiens is a part of nature), it seems arbitrary to actively conserve "native" species that arrived in a place by some other-than-human means and to eradicate those "exotics" that got there by deliberate or inadvertent human agency. An alternative conception of what to conserve is biological integrity, which is defined in terms of a "natural" biotic community, free of human influence. But again, from the post-Darwin perspective, it seems arbitrary to single out human influence as opposed to, say, beaver influence or coyote influence, as that which compromises the "naturalness" of an area.[16]

The Boundary Between the Human and Natural Worlds Drawn by the Temporal Scales of Biological and Cultural Evolution

Paul Angermeier reviews the conservation-biology literature on the confounding and paradoxical boundary between what Fritzell defiantly calls "man" and nature, and points the way out of this post-Darwin fly-bottle. At first hearing, Angermeier seems to affirm only a small variation on one of the age-old essentialist themes: "Human activity becomes unnatural when it involves technology."[17] Homo sapiens is not the uniquely rational animal; we are, rather, the uniquely technological animal, a variation on the tool-using essence.

But the reason Angermeier gives for technology being the boundary marker (and maker) separating the human and the natural worlds is the temporal scale of the *"evolution"* of technology. "Human activities that exceed our genetically evolved—as opposed to *culturally evolved*—abilities are unnatural," he avers.[18] At bottom, as Angermeier ultimately recognizes, what draws the boundary between the human and natural worlds is not technology per se, but the disparity between the temporal scale of cultural evolution and genetic evolution. Technology is only one component of culture, but the one Angermeier singles out as boundary-defining because that component of culture has the most impact on nature: "Humans are cultural as well as biological animals. For conservation, the most important outgrowth of culture is technology, with which we transform nature.... Because technological [and more generally cultural] evolution is much more rapid than genetic evolution, we transform ecosystems faster than other biota can adapt."[19]

What is cultural *evolution*? Culture appears to "evolve" in the sense that simpler and more rudimentary cultural items are succeeded by more complex and sophisticated cultural items. For example, a sharpened stick used for spearing is succeeded by a stone- or antler-tipped spear, which is succeeded in turn by a simple bow and stone-tipped arrow, which is succeeded by a cross bow and an iron-tipped arrow, which is succeeded by a blunderbuss, which is succeeded by a rifle. Or, a birch-bark/hot-stones cooking vessel is succeeded by a clay pot, which is succeeded by an iron kettle, which is succeeded by a Dutch oven, which is succeeded by a gas oven, which is succeeded by a microwave oven. Or, pictographs are succeeded by hieroglyphs, which are succeeded by alphabets; and clay tablets are succeeded by papyrus scrolls, which are succeeded by paper manuscripts, which are succeeded by printed books, which are succeeded by digital or virtual docs. Or mythopoeia is succeeded by speculative natural philosophy, which is succeeded by modern hypothetico-deductive-experimental Newtonian science, which is succeeded by postmodern statistical-probabilistic Heisenbergian science. One may be reluctant to call such lines of succession in cultural items "progress" because the abandonment of simpler, more rudimentary cultural items in favor of more complex and sophisticated cultural items may come at a high cost in personal satisfaction and happiness and social stability. But whether progressive or not, culture appears to evolve in this way.

Cultural evolution is much faster than genetic evolution because it is Larmackian, not Darwinian. Jean-Baptiste Lamarck believed that species evolved through the transmission to offspring of characteristics acquired by parents. For example, if a man were a right-handed blacksmith and had acquired an enlarged right arm as a result of repeated hammering with that arm, while the other remained mostly at rest holding the molten iron, then half his children on average would be born with a larger right than left arm.[20] That notion has long been abandoned in biological evolution in favor of Darwin's scenario of chance variation (later to become genetic mutation) and natural selection among genotypes. But the inheritance by subsequent generations of novelties acquired by previous generations is indeed the mode of cultural evolution. And Lamarckian evolution is much, much faster than Darwinian evolution.

To illustrate graphically and mythically the difference in temporal scale between biological and cultural evolution, suppose you are *Ursa*, the bear diva, and you are seized with a desire to live in the Arctic and eat the delicious flesh and luscious fat of seals. To satisfy your desire to be a polar bear, you must increase your body size, modify your digestive tract, change the color of your fur, modify your hibernation cycle, and make a thousand other anatomical, physiological, and behavioral adaptations, each of which must await the chance of a favorable genetic mutation, a process requiring more than a million years to complete. Then these changes must accumulate in the gene pool of a population and become stabilized. If you are a Homo sapien and wish to adapt to life in the Arctic and consume its rich protein resources, you may thoughtfully and deliberately modify your existing material and cognitive culture—you may transform a spear into a harpoon, a pole and bark lodge into an igloo; you may learn the difference between the ways of grizzly and polar bears, and the difference between the dispositions of the spirits in the taiga forest and those of the drift ice. If these and a thousand other such cultural adaptations work and you are successful in your Arctic enterprise, you may pass them on not only to your own offspring, but to the offspring of all the members of your band, a process completed in just a few human generations.

The post-Darwin human-natural paradox can now be stated more clearly. The genus Homo, from a Darwinian evolutionary point of view, is—with four other genuses: Pongo, Hylobates, Gorilla, and Pan—an anthropoid ape. Unlike the other apes, Homo sapiens evolved culture, or perhaps better—allowing that other species may have some rudimentary culture—what Charles Lumsden and Ed Wilson call "euculture."[21] Lumsden and Wilson argue that, having reached a certain takeoff point, human cultural and genetic evolution were linked in a positive feedback loop. Culture became in effect a selective environment in which Homo evolved. Culturally adept individuals—those who could learn, use, and store information and communicate it to their offspring and other close kin—reproduced more successfully than those who were less so. The greater an individual's capacity for acquiring cultural skills, the greater their inclusive fitness.

The principle organ of human culture, the brain, evolved rapidly (on the scale of biological evolution), tripling in size over just three million years. The Australopithecine cranial capacity was 400–500 cubic centimeters, comparable to that of the modern chimpanzee and gorilla; that of Homo erectus was about 1,000 cubic centimeters; that of Homo neanderthalus about 1,500 cubic centimeters; and that of Homo sapiens about 2,000 cubic centimeters. In sum: Homo sapiens is an anthropoid ape, evolved, as any other species, by the general neo-Darwinian process of random genetic mutation and natural selection. Human culture also evolved by the general neo-Darwinian process of random genetic mutation and natural selection, but itself became part of the selective environment in which Homo evolved. That is, both human beings and human culture are a part of and product of nature. But because the temporal scale of cultural evolution, vis-à-vis biological evolution, is so disparate, euculture has propelled Homo sapiens out of nature. The disparity, moreover, between the human and natural worlds is increasing precisely because the rate of cultural evolution is increasing. In other words, cultural evolution occurs at a faster temporal scale than biological evolution; but not only that, while the rate of biological evolution remains constant, the rate of cultural evolution increases dramatically. Perhaps that's one reason why we tend to think that premodern iterations of Homo sapiens are closer to nature than the modern and now postmodern ones.

Back now to the contradiction Peter Fritzell finds in Leopold's land ethic. Leopold seems to think that not only are we a part of nature, but, for a land ethic to emerge, we must *acknowledge* that we are indeed a part of nature—we must view or see ourselves as a part of nature. And, from the evolutionary point of view assumed here, we are indeed a part of nature. But, as Fritzell points out, to have an evolutionary point of view is—as we may now say—a "meme," an item of culture. And to the extent that we are eucultural animals, we are set apart from nature. Now, with the help of this analysis of the linkage between biological and cultural evolution and the disparity between the temporal scales at which adaptive processes proceed in either scale, that we are both a part of nature and set apart from it ceases to be a contradiction; indeed it ceases to be even paradoxical.

How Drawing the Boundary Between the Human and Natural Worlds in Terms of the Temporal Scales of Biological and Cultural Evolution Establishes Defensible Norms for Environmental Ethics

Understanding the boundary between the human and natural worlds to be drawn by the disparity between the temporal scales of cultural and biological evolution is useful not only for purposes of defending conservation biology's classic naturalness norm; it is useful for establishing norms for environmental ethics as well (now that the penumbra of contradiction and paradox surrounding environmental ethics has been dispelled).

At this late date on the Lamarkian temporal scale of cultural evolution, every cubic centimeter of the biosphere has suffered measurable effects of human culture, but some, obviously, much more than others. Naturalness admits of degrees and conservation biologists can aim to preserve or restore relatively or comparatively natural areas, even though purely natural areas (pristine areas) are nowhere to be found.

One concern of environmental ethics is to suggest and justify principles or precepts to govern the human treatment of nature. As noted, the idea that nature is in a state of static equilibrium has long been abandoned in ecology. Contemporary ecology has shifted from the "balance-of-nature" paradigm to the "flux-of-nature" paradigm. And as here amply summarized, nature constantly changes at multiple temporal scales. So why should some human changes imposed on nature be ethically problematic? If volcanoes blow off mountain tops and raze the forests on the slopes below, fill valleys with debris, and pollute streams, what's wrong with strip mines that have comparable destructive effects? If fires periodically rage through forests, burning them to the ground, what's wrong with clear-cuts that have comparable destructive effects? If hurricanes erode ocean-front sand dunes, what's wrong with beachside cottage developments that have comparable destructive effects? Because Homo sapiens is a part of nature, we cannot simply answer, without further consideration, that such effects are anthropogenic, as if they were caused by a species of space alien.

What renders strip mines, clear-cuts, and beach developments unnatural is not that they are anthropogenic—for, biologically speaking, Homo sapiens is as natural a species as any other—but that they occur at temporal and spatial scales that were unprecedented in nature until nature itself evolved another mode (the Lamarckian mode) of evolution: cultural evolution. Steward Pickett and Richard Ostfeld note that "for all its scientific intrigue and poetic beauty, the flux of nature is a dangerous metaphor. The metaphor and the underlying ecological paradigm may suggest to the thoughtless and greedy that since flux is a fundamental part of the natural world, any human-caused flux is justifiable. Such an inference is wrong because the flux in the natural world has severe limits."[22] Pickett and Ostfeld go on to identify the general kinds of radical constraints on natural change (as we may now call it without confusion about whether anthropogenic change is a subset of natural change). And one of these is the incursion of the temporal scale of cultural change on that of biological (or Darwinian) evolutionary change. "In general terms, these limits of natural flux are functional, historical, and evolutionary.... Problematic human changes or fluxes are those that are beyond the limits of physiology to tolerate, history to be prepared for, or evolution to react to. Two characteristics of human-induced flux would suggest that it would be excessive: fast rate and large spatial extent."[23]

So, back to our examples of natural versus anthropogenic environmental changes. Catastrophic volcanic eruptions, such as that which literally blew the top off Mount St. Helens in 1980 occur in widely scattered frequencies and localities. However, "mountaintop removal" coal mining in West Virginia occurs at such a huge scale that no figures are available for its spatial extent and temporal frequency—an anthropogenic flux now so intensive that a mountainous region of North America is well on its way to becoming a region of tabletop plateaus. Clear-cuts could be ethically justified, on the other hand, if their spatial and temporal scales approached those of nature fire regimes or insect-infestation cycles that afflict forests on the ecological temporal scale. Cottage developments on beachfront property could also be justified if they were mindfully fitted, at comparable scales, into a matrix otherwise subject only to natural disturbance regimes.

The environmental immorality of the most disturbing of all anthropogenic changes imposed on nature—the sixth great extinction even in the 3.5 billion year biography of planet Earth—can only be persuasively articulated as the boundary violation of the evolutionary temporal scale by the cultural. Species extinction is a natural process, but the scale at which it normally occurs—the "background rate" of extinction—is slower than speciation. We can know that a priori, for if the rate of extinction were not slower than the rate of speciation, Late Pleistocene biodiversity—to the tune of five to thirty million coexisting species, when human beings inherited the Earth—could not have accumulated. Natural species extinction thus occurs normally on the evolutionary temporal scale, but at a rate slower than the rate of speciation. Anthropogenic species extinction now occurs at a boundary-busting rate many orders of magnitude faster than speciation. There is nothing environmentally unethical about anthropogenic species extinction per se. What is wrong with the current episode of abrupt, mass, anthropogenic species extinction is—and is only—the temporal scale on which it is occurring.

Notes

1. For this historical background I consulted Liddel and Scott's *Greek-English Lexicon*, 7th ed. Oxford: The Clarendon Press, 1964; G. S. Kirk and J. E. Raven, *The Presocratic Philosophers: A Critical History with a Selection of Texts*. Cambridge: Cambridge University Press, 1962; W. K. C. Guthrie, *A History of Greek Philosophy*, Vol. 1: *The Earlier Presocratics and the Pythagoreans*. Cambridge: Cambridge University

Press, 1962; and Richard D. McKirihan, Jr., *Philosophy Before Socrates: An Introduction with Texts and Commentary.* Indianapolis: Hackett Publishing Co., 1994.

2. C. S. Holling, "Cross-scale Morphology, Geometry, and Dynamics of Ecosystems," *Ecological Monographs* 62 (1992): 447–502, 449.

3. Ibid.

4. Ibid., 449–450.

5. For this discussion of vegetative processes I have consulted Michael J. Crawley, ed., *Plant Ecology*, 2nd ed. London: Blackwell Science, 1997.

6. For this discussion of ecological processes I consulted Crawley, ed., *Plant Ecology*; S. T. A. Pickett and P. S. White, *The Ecology of Natural Disturbance and Patch Dynamics.* San Diego: Academic Press, 1985; and R. Moss, A. Watson, and J. Ollason, *Animal Population Dynamics.* New York: Chapman and Hall, 1982.

7. For this discussion of evolutionary temporal scale, I consulted E. O. Wilson, *The Diversity of Life.* Cambridge: Harvard University/Belknap Press, 1992, and Ralf Hennemann, *Sharks and Rays: Elasmobranch Guide to the World.* Frankfurt: Unterwasserachiv, 2001.

8. For this discussion of plate tectonics and the rock cycle, I consulted B. C. Birchfield, R. J. Foster, E. A. Keller, W. M. Melhorn, D. G. Brookins, L. W. Mintz, and H. V. Thurman, *Physical Geology: The Structures and Processes of the Earth.* Columbus: Charles E. Merrill, 1982, and F. Press and R. Siever, *Understanding Earth.* New York: W. H. Freeman and Co., 1994.

9. D. B. Botkin and E. A. Keller, *Environmental Science: Earth as a Living Planet,* 2nd ed. New York: John Wiley & Sons, 1998, 56.

10. For this discussion of the climatic temporal scale I consulted W. F. Ruddiman, *Earth's Climate: Past and Future.* New York: W. H. Freeman & Co., 2001, and R. G. Barry and R. J. Chorley, *Atmosphere, Weather, and Climate,* 6th ed. New York: Routledge, 1992.

11. For this discussion of interactions at the boundaries between temporal scales I consulted R. V. O'Neill, D. L. DeAngelis, J. B. Waide, and T. F. H. Allen, *A Hierarchical Concept of Ecosystems.* Princeton: Princeton University Press, 1986.

12. Aldo Leopold, *A Sand County Almanac and Sketches Here and There.* New York: Oxford University Press, 1949, 109.

13. Ibid., 204.

14. Ibid., 203.

15. Peter Fritzell, "The Conflicts of Ecological Conscience," in J. B. Callicott, ed., *Companion to A Sand County Almanac.* Madison: University of Wisconsin Press, 1987, 128–153, 14.

16. This discussion of norms for conservation biology is informed by P. L. Angermeier and J. L. Karr, "Biological Integrity versus Biological Diversity as Policy Directives," *BioScience* 44 (2000): 690–697.

17. Paul L. Angermeier, "The Natural Imperative for Biological Conservation," *Conservation Biology* 14 (2000): 373–381, 374.

18. Ibid.

19. Ibid., 375.

20. Richard W. Burkhardt, *The Spirit of System: Lamarck and Evolutionary Biology*. Cambridge: Harvard University Press, 1977.

21. Charles L. Lumsden and Edward O. Wilson, *Genes, Mind, and Culture: The Coevolutionary Process*. Cambridge: Harvard University Press, 1981.

22. S. T. A. Pickett and R. S. Ostfeld, "The Shifting Paradigm in Ecology," in R. L. Knight and S. F. Bates, eds., *A New Century of Natural Resources Management*. Washington: Island Press, 1995, 273.

23. Ibid., 274.

Chapter 3

The Ethical Boundaries of Animal Biotechnology: Descartes, Spinoza, and Darwin

Strachan Donnelley

We have been asked to explore the taxonomy of boundaries. Philosophers have long struggled with the problems of boundaries, especially conceptual boundaries that block or distort philosophic insight and a faithful rendering or interpretation of human experience. What may seem to some a highly esoteric exercise, in fact, directly or indirectly, has central practical import. Fundamental philosophic worldviews are established that help to articulate our understanding of the world, including the full range of our moral responsibilities to humans and the rest of nature.

A particularly instructive case is the philosophic struggles or differences between Descartes, Spinoza, and Darwin, which sheds light on an emerging philosophic worldview that may help us to think through a new and complex moral issue: animal biotechnology, the genetic manipulation of animals for human purposes.

Let us begin by noting exactly what is at stake, namely, our intertwined ethical and civic responsibilities to the future of humans, animals, and nature—boundary and transboundary questions par excellence. Beyond familiar moral issues of human and animal sentience (pain, suffering, individual life goals or plans), philosophy and ethics must address more elusive questions about the moral significance of the biological and behavioral integrity of individual organisms (including animals) and the integrity and flourishing of natural populations and communities of organisms, particular ecosystems, and natural evolutionary and ecosystemic processes. Such systematic considerations are necessary for finally deciding what we ought and ought not to do with respect to the human use of animals and nature. Again, rubbing together the philosophic sticks of Descartes, Spinoza, and Darwin may prove particularly useful in addressing these far-reaching and increasingly urgent issues.

Cartesian Cosmology

With an eye toward Spinoza, I start with Descartes and his philosophic assertions of the ontological primacy of "substance" and of a multiplicity of substances. Descartes fatefully defines a substance as "that which needs nothing else in order to exist." The paradigmatic instance of substance is God, the ultimate powerful and self-sufficient one. This definition ultimately resigns all substances, whether of the same or different natures, to merely unessential or *external relations*. Substances do not, and ontologically cannot, essentially affect or matter to one another. Whatever relations might hold between them must be adventitious, contingent, and have no bearing on their very being or essence.

Descartes recognizes three types of substance, each with its own unsharable attribute: God, mental substance, and physical or material substance. Whatever the relation of the other two types of substance to God, *res cognitans* and *res extensa* are essentially independent of each other and have radically different attributes or characters: thinking, including willing, and mere extension (extensiveness in width, breadth, and depth) to which motion must be added to gain natural, physical dynamism.

This is Descartes's famous metaphysical dualism, which banished all subjective activity and agency from the natural realm and left nature free for the mechanistic causal explanations of the new mathematical science, which was primarily concerned with inorganic nature. Yet this metaphysical service to emerging scientific materialism was bought at a heavy philosophic price—a price particularly germane to our ethical interests in humans, animals, and nature.

Relationships

First is the problem of relationship or connectedness. Minds (*res cogitans*) and bodies (*res extensa*) share absolutely nothing in common, and it is philosophically unintelligible how there can be any direct connection, relation, or communication between the two realms of substance. Against the primary evidence of our experience, it becomes impossible to comprehend how a thinking subject can direct bodily action and can encounter, know, and think about the external, radically other realm of nature. Descartes's own doctrine of innate, clear and distinct, and thus certain ideas about nature abroad (established in us by God) strikes us moderns as an epistemological deus ex machina.

Nor is the relational problem merely between the two realms of mind and matter. Each realm itself is infected with philosophic unintelligibility. If we suppose a plurality of mental substances or human minds, how do they get over to and communicate with one another without some use of the radically other physical realm as a medium of communication (e.g., written symbol or sound)? How can they break out of a splendid, isolated independence and an egocentric, solipsistic predicament? How do they influence one another's subjective life, or even be assured of each other's existence?

On the physical side of the metaphysical divide, we encounter similar problems. We may opt for nature as a single physical, extended substance and avoid the problem of external relations to other physical entities altogether. But then, within the mechanistically determined plenum, what accounts for the systematically related articulations of nature, most pointedly complexly organized, bodily organisms and the communities and ecosystems that they together fashion?

If we opt for the atomistic route and an indefinite multiplicity of extended substances, we are no better off. For each extended atom or concatenation of atoms does not require any other physical entities in order to exist. Nor will their physical characteristics, singly or in aggregate, essentially require or be related to one another. Again, all natural order would be radically contingent. Blindly running, externally related billiards balls would have to foot the explanatory bill for the historically or temporally evolved, internally articulated order of nature, including the terrestrial realm of animate life.

All this is ontologically implausible. The price of philosophic unintelligibility is just too high. Descartes proposes a philosophy of spiritual, intellectually angelic minds and dead, inorganic, mechanistic nature in which biological organisms and the realm of animate life do not fit. For it is precisely with biological organisms, including our human bodily selves, that there can be no radical split between the realms of mind and organic body. The organic subject or self acts into the world with and through its body, and the living subject is aboriginally introduced to the world via the body's sensory functionings. Further, the particular definite character and capacities of the living organism—physical, physiological, behavioral, and psychic— are coordinated and teleologically structured for the sake of the lively existence of the integral organism itself. Still further, these characters and capacities speak for or reflect the fundamental features of the natural world out of which the organism has evolved. In short, there is a dynamic, if imperfect fit between organism and environing world. In all dimensions of organic, particularly animal, existence, worldly relatedness and dynamic communicative interaction are as essential as psychophysical individuality or selfhood. All this is lost in Descartes's dualistic substance philosophy.

Value and Nature

It is only left to draw out the value implications of Descartes's philosophic picture. By metaphysical fiat, all subjective agency and purpose are written out of nature, leaving a realm of blind, causally efficient forces. This by definition is a valueless realm, since nothing of worth can be internally aimed at and achieved, and nothing can be benefited or harmed. Nature at best can only have a pragmatic or instrumental value for an external, observing, thinking subject. But even this value is philosophically problematic, for the thinking subject has essentially nothing to do with nature existentially or practically and is thus only inexplicably and unintelligibly related to it.

In short, Cartesian valuation, whether ethical, aesthetic, or religious, must be decidedly anatural and human and/or God centered. It cannot directly value the individual living integrity and complex integral forms, psychic capacities, and concrete activities found in and between living organisms, for they have none. All is blind, purposeless necessity. All organisms, plant and animal, and the ecosystems that they help to constitute are but natural mechanisms, contingently arrived at or fashioned by God, to be used as human prudence dictates. Logically Cartesian eyes are and must be blind to natural values. There can be no authentic Cartesian ethics of the natural animate world.

Spinoza's Cosmology

With respect to organic and humanly organic life, Cartesian modes of thought leave us in a philosophic cul de sac. A chief stumbling block is the conception of substance as "that which needs nothing other in order to exist." With such a definition, organisms logically and ontologically *cannot* be substances, for by their fundamental mode of metabolic existence, they essentially require the physical, natural other in order to be. For organisms, individuality (and thus any particular integrity or independence) and relatedness to the world must be equiprimordial.

To conceive philosophically these fundamental and interconnected themes of individuality and relatedness, it is necessary to move decisively beyond Cartesian modes of thought. This was clearly seen by Spinoza. Spinoza abandoned Descartes's dualism and assertion of multiple substances in favor of a radical monism. There can be only one being or entity that requires "nothing else in order to exist." This is God or nature (*Deus sive*

Natura) as a whole, *causa sui* (cause of itself), acting out of the necessity of its own dynamic being, characterized by an infinity of attributes of which we humans know only thought and extension.

With this one substance cosmology, Spinoza philosophically allows for a fundamental reconceptualization of organic, including human, life. Biological organisms and human beings are no longer respectively conceived as natural mechanisms and anatural, radically independent thinking subjects. Rather, both are considered finite modifications or internal articulations of the unbounded active being of nature (God), with a "modal" rather than "substantial" existence.

With the conception of modal existence, Spinoza is philosophically able to combine the themes of individuality and relatedness. The essence or fundamental character of each finite mode (concrete natural entity) is an *endeavor to persevere in its own individual being*. This is its "conatus," the basis of its individuality, which necessarily implies dynamism, activity, and recurrent effort. Since by necessity all finite modes are causally interconnected and directly or indirectly interactive in the one infinite system of nature, conative existence can only be pursued in essential relation to the worldly others. Concrete individuality must be an ongoing, dynamic affair, won *within* a relatedness to the world. In short, an organism's particular and individual existence is *internally related* to the world.

This conception of a "conative mode" systematically opens up new and important possibilities for philosophic explanation and intelligibility. Spinoza conceives organisms, human and other, as ontologically one, conceivable alternatively under the attributes of thought and extension. This is Spinoza's famous "psychophysical parallelism." What is conceived to happen causally under the attribute of extension (body) also and correlatively happens causally under the attribute of thought (idea of body), for ontologically they are one and the same happening. The conceptions of mind and body are an intellectual distinction that involves no "real" difference.

Consider the philosophic gains. Whether conceived under Spinoza's strict natural necessity or the doctrine of more historically contingent neo-Darwinian evolution, organisms are understood to be naturally organized for worldly experience and action that essentially serve the conative endeavors of the organic individual. This must mean that real potentialities for organic individuality, activity, and experience are woven into the very fabric of nature. Moreover, there can be varying, naturally constituted levels of organic complexity and organization, yielding conative individuals (human and animal) with varying capacities for worldly activity and experience, for the twin powers of acting on and "undergoing" or "suffering" the world.

We can now philosophically understand why we and other organisms are necessarily worldly actors and sufferers, with psychic and bodily capacities attuned to the world. We must be in order to be. Furthermore it is a decided natural fortune that all organisms, most emphatically ourselves, have been forged in the crucible of worldly, natural, dynamic existence. Our perceptual and sensory capacities are naturally constituted to reveal more or less adequately the realized order of the world. The objective, worldly other importantly determines the subjective or individual psychic "inner," for nature's finite conative entities mutually and interactively (indirectly, if not directly) determine one another's character, physical and mental, and the overall realized order of nature.

In short, by natural necessity, we in our own particular and finite human way are attuned to the natural world. In pursuing our own conative and metabolic mode of existence, we perceive in our limited and imperfect human fashion the order of the world and the interactions among its constituents, and we in turn bodily act on or with them. We as complex organisms take account of or "know" the world, which includes its own natural values and goodness. These "natural values" primarily are nature's organisms themselves: the various concrete forms and instances of dynamic, conative, and active existence that are interdependently fashioned in the system of nature. They are concrete values in the sense that they essentially matter to themselves, to the modal others, if not also to the natural whole. (This is a fundamental assertion held in common by Jonas, A. N. Whitehead, and other philosophic naturalists.)

It is within this wider animate and worldly context of existence that we creatively fashion our peculiarly human world: human communities (the human polis) and the complex cultural realms of humanly significant, meaningful, symbolic existence and activity. These human endeavors include our political, intellectual, artistic, spiritual, and ethical creations. Yet it is important to keep in mind that the decidedly human world is possible only on the basis of our natural status as organic, conative creatures within the world. The human world is complexly parasitic on the natural, animate world. The order of the natural world must be there prior to our taking account of and creatively acting on it. Higher human capacities of imagination, thought, and purposeful, bodily action presuppose powers of abstraction, of lifting "forms" out of their concrete contexts. These achievements fundamentally depend on both the bodily organism's perceptual capacities and a definite order within the given natural world, no matter how freely we may vary these abstracted forms in consequent mental activity. (Humans *cannot* be divorced from nature. So forcibly argues Hans Jonas, among others.) Moreover, for these abstracted and freely varied forms to become genuine symbols, carrying their own weight of human significance and meaning in all their infinite variation, we by necessity must be conative creative creatures living within a dynamic and efficacious world. Things can have felt meaning or significance for us primarily because our own efforts and the world existentially matter to us and matter because of a dynamic and complex interrelatedness. On ourselves and the world coequally depend our very individual existence. In conative being, naturally organic and cultural existence meet and intricately interweave, and perhaps not exclusively on a human level. Other forms of organic, animal life seemingly lead their own forms of "signal," if not fully symbolic and meaningful existence.

Darwinian Cosmology

The contested terrain of philosophic cosmology has not been subdued by Spinoza, despite his self-proclaimed true (if not the best) philosophy. Other contestants have entered the field. Of particular interest to us is the 1859 publication of Charles Darwin's *The Origin of Species*.

According to Ernst Mayr, the recently deceased dean of evolutionary biology and philosophy, Darwin's complex theory of evolution (the historical evolution of all life from a single origin via genetic and behavioral variation and natural selection) constitutes the most profound revolution in the whole history of Western science and philosophy. Central assumptions and presuppositions of Western thought have been undermined. Mayr fastens particular attention on cosmic teleology, physicalist determinism, and essentialism or typological thinking. I will briefly examine his argument in relation to Spinoza's philosophic advances over Descartes.

The first western philosophic pillar to fall at the hands of Darwinian biology is cosmic teleology—nature as the Grand Design of the Grand Designer (rational or willful). Rather, nature engenders its own forms and order in passing, whether gnomic, organismal, ecosystemic, bioregional, or ecospheric. Though Spinoza also denied cosmic teleology or purpose—God's very being determines all—the rational character of Spinoza's God or nature is superceded as far as any direct implication in earthly becoming. The move to an arational nature constitutes a major shift in philosophic cosmology.

The second Western pillar to fall is physicalist, Newtonian determinism. We met these billiard-balls-in-motion in our discussion of Descartes. Such straightforward, if many stranded, determinism is out. Rather natural organic entities and events are determined by multiscaled temporal and spatial causes, with historical chance and contingency playing a significant role. Moreover, Mayr points to *ultimate*, as well as *proximate* or psychical-chemical causes. Ultimate causes are evolutionary causes, issuing from the historical, worldly evolution of genomes via variation and selection. Genes and genetic material, "informational matter," were unknown to hard-nosed Newtonians and Spinoza. Here is an important new ingredient to be woven into philosophic cosmology and ethical reflection.

The third Western pillar to fall is perhaps the most disruptive of all to traditional cosmologies. This is essentialist or typological thinking. Species types or essential forms—dog, rose, human being—with all individuals of a particular species essentially the same and only accidentally (contingently) different, are gone. Rather, with rare exceptions, all individual organisms are genetically and phenotypically unique, different from one another. Such individual difference and uniqueness hold as well for populations and communities of organisms, ecosystems, bioregions, and more. Further, these differences and variations are not merely adventitious. They are the warp and woof of evolutionary life. Without such requisite variation, natural selection would have nothing to select. Adaptive and adapted life would come to a standstill, or rather would never get going. This natural diversity, plurality, and messiness subvert traditional philosophic cosmologies, including that of Spinoza.

With respect to contrasting Spinoza and Darwin, I want to fasten on two key conceptions: causation and individuality (human and other).

As noted, Newtonian deterministic causation is out. "Orchestral causation" is in. Mayr would have us imagine an orchestral performance. Who or what is the cause: the orchestral score, the composer, the conductor, the respective members of the orchestra, the audience (variously musically capacitated), the orchestral setting (the orchestra hall)? There are no single, "necessary and sufficient" causes. The resultant symphonic music *emerges* from the interactions of all these, and no doubt other, factors. The musical emergent is unique and historically contingent. Change any of the factors and interactors and the results would be unpredictably different. Spinoza's modal realm was similarly interactionist, but the stakes were not so high or cosmologically significant. Historical interactionism constitutes Darwinian reality, its very being and becoming. There is no permanent and eternal Spinozistic god atemporally engendering and taking note of the derivative and passing modal play.

Orchestral causation leads directly to the conception of an individual organism, human or other. Spinoza's conative individuals are co-constituted and co-determined by their relatedness to the (finite) modal others. Darwinian modes of explanation only deepen this line of thought. Darwinian individuals owe their very being and character to the natural, historical world. Not only do they derive their individual genomes from an immemorial, 3.8-billion-year past, the genome must interact with its particular cellular, somatic, and ecosystemic (worldly) environment if the unique phenotypic individual is to emerge at all. Indeed, the individual organism is an *emergent* and is ongoingly so until it dies. This holds for human individuals or selves as well. Particular cultural and social settings are only added to the historical, worldly mix. Emergent selves and orchestral causation go hand and hand. They mutually require one another in Darwinian philosophic cosmology.

These shifts in cosmological explanation directly open up the question and philosophic interpretation of human ethical concern. Given an eternal God or nature and the necessary vicissitudes of modal existence, Spinoza's *Ethics* marches inexorably to issues of individual freedom, the humanly good life, and the politically liberal institutions required for such humanly conative existence. The rest of the conative realm is, and could be, left to its own devices given the atemporal and invulnerable eternity of divine/natural activity.

In Darwinian cosmology, locus of ultimate ethical concern by necessity shifts. Darwin's and our ultimate locus of concern is the historical realm (past, present, and future) of natural and cultural becoming itself. (Here is a realm of being/becoming truly worthy of ultimate, spiritual respect, awe, and gratitude.) In this realm, nothing is assured or can be taken for granted. All life alike and on all levels is finite, mortal, and vulnerable, especially now to our human misadventures. Thus, moral responsibility to the human and natural future must take ethical pride of place. This responsibility must become a chief ingredient of the humanly good life and the exercise of individual freedom. Or so Hans Jonas argues forcibly and persuasively in *The Imperative of Responsibility*. The course of my argument is heavily indebted to the philosophic writings of Hans Jonas, especially *The Imperative of Responsibility*, *The Phenomenon of Life*, and *Philosophical Essays*

Spinoza, Darwin, and Animal Biotechnology

With this Darwinian update or transformation of Spinoza's cosmological worldview, we are ready to tackle directly the practical moral issues of animal biotechnology. We must bypass certain knotty philosophic issues that would only take us off course. These include the nature of philosophic worldviews or cosmologies. (They are not certain knowledge or proofs, but only speculative interpretations of the world.) There is also the relation of philosophic worldviews to moral judgments. (Moral judgments are informed by, but not logically derived from, philosophic schemes of thought.) What we need to remember here is the fundamental interconnection and interactions of humans, animals, and nature emphasized by Darwinian worldviews. But on to animal biotechnology and the various purposes and uses of genetic manipulation of animals.

Straight off we should note that from a Darwinian naturalistic perspective, biotechnological intervention per se (genetic or otherwise) is not the primary locus of ethical concern. The naturally conative realm, irrespective of human interventions, is continually, dynamically, and internally changing its forms (genetic and other) and the methods of this change seem prima facie morally insignificant. Nor does the change or mixing of individual organisms' genetic material seem ethically important, for genes are not the point, the end, or the final good of the animate realm. (All organisms share genetic material in any case.) What manifestly is ethically significant is the effect or consequences of biotechnological manipulations and interventions on the organic, conative integrity and complexity (richness) of individuals, communities, and ecosystems. Integrity and impoverishment, in their many dimensions, are the key ethical issues. For example, we should be ethically chary of interventions, biotechnological or other, that might undermine integrated and complex human capacities of experience, thought, and action (moral and other) and threaten the fundamental terms of our individual and communal worldly existence. Similarly we should be ethically wary of introducing organisms, biotechnologically engendered or not, into new or foreign environments, with possible systemic and seriously disruptive effects on the well-being of ecosystems, animal populations, and the naturally evolved, conative capacities and behavior of individual organisms. Examples include the introduction of chemical fertilizers, "natural pesticides," and other microorganisms into local environments; the unintentional introduction of exotic species in aquatic systems (e.g., zebra mussels in the Great Lakes); and the release of transgenic or pen-raised hatchery fish into wild river systems.

A first general ethical injunction (gratefully borrowed from Western moral tradition, still appropriate in a new worldview of humans and nature) is to do no harm to animate individuals and communities without morally good reasons (i.e., other conflicting values or obligations). A second, more stringent ethical injunction is to do no significant, "out of the naturally ordinary," especially irreparable harm to the natural animate realm for any reason. (This is a moral imperative especially emphasized by Jonas in his post-Darwin *Imperative of Responsibility*.) The difficulty, of course, comes with moving from the general injunctions to concrete particular cases, in ethically and scientifically agreeing on what are good reasons and significant harms, and in making trade-offs between harms and benefits that involve disparate values and/or levels of animate life. For example, we may or may not agree that it is ethically legitimate to harm the conative integrity of a certain environmentally isolated or benign strain of mice (the transgenically fashioned Onco or Alzheimer's mouse) for the sake of medical or basic biological research. (Here is a mixture of preventing significant human harm and promoting a particularly human good [scientific knowledge] at the expense of animal wellbeing.) Would we make the same ethical trade-off if the research animals were chimpanzees or gorillas, which are threatened species with richer conative capacities and more elaborate social relations, or if the research were in the wild and involved major disruptions to animal populations or communities and threats to the ongoing viability of a richly differentiated ecosystem? The practical answers would depend on how we ethically value human, animal, and ongoing ecosystemic well-being relative to one another and on how we ought to juxtapose the avoidance of harms, with the prevention of long-term, systemic harms taking ethical precedence over harms with which we can readily live.

A relevant case study ripe for such ethical reflection is xenotransplantation and the potential use of transgenic pigs as organ donors for humans undergoing organ failure. Pigs have a porcine retrovirus that theoretically could find its way into human genomes and mutate into a lethal form that could cause systemic havoc as with the AIDS epidemic. The odds of such a catastrophe is literally incalculable, but should we put human (and perhaps other) communities seriously at risk, now or in the distant future, for the sake of presently distressed human individuals, no matter what the odds?

These examples involve conflicting and perhaps incommensurate values and ethical obligations. In each case the several questions of ethical responsibility or obligation must be raised and decided together, perhaps without any final certitude, within the context of our moral landscape and general ethical imperatives. We must cross boundaries of specific ethical concern. We necessarily must practice the fallible art of moral judgment (moral ecology: considering all moral values and responsibilities together), for there are no simple, clear-cut, and straightforward ethical answers to these cases provided by philosophic cosmology or other philosophic resources.

Nevertheless there is a crucial side effect of biotechnology that easily escapes notice or is brushed aside and is ethically difficult to assess. This is the moral effect of biotechnology, animal or other, on the meaning, significance, and symbolic systems of human life: on aesthetic, spiritual, and ethical values and their symbolic cultural expressions, "bounded" dimensions of human life that ought to be taken into account.

Take the case of introducing transgenic or pen-raised hatchery fish, for example, trout or salmon, into wild streams or rivers for the sake of sport fishing (which of course raises its own ethical issues). What if, against our best evidence (we know better), the introduction has no adverse effects on the aquatic ecosystems and native populations of species life: fish, insect, and other? Would it make any ethical difference to human life? Arguably it would. Purpose-bred or genetically engineered animals become dominantly characterized for us by their instrumental purpose, as does a hammer. Here the purpose might be the trout's or salmon's size, fighting ability, or eagerness in taking an angler's lure. Whichever, the fish have been initially and intentionally removed from a natural and historically deep context of ethological, evolutionary, and ecological processes and humanly impounded, isolated, and manipulated. In our experience, they are domesticated and suffer a symbolic reduction to their humanly conceived purpose. They no longer have the meaning or significance of a wild trout or salmon in its native aquatic setting.

When they are reintroduced into a natural setting, they tend to undermine its wild symbolic valence for humans, if not also devastate the aquatic ecosystem itself.

By this "symbolic valence, meaning, or significance," I mean animals, ecosystems, and nature experienced by us humans as an inexhaustible resource of multiple values and valuations: aesthetic, spiritual, ethical, intellectual, and practical. (Here is where a Darwinian philosophic perspective importantly counts.) Most primordially, there are the protean, self-renewing, form-engendering, if vulnerable powers of nature itself, expressible and expressed in innumerable symbolic forms and systems artistic, philosophic, religious, scientific, and other. This is nature, for us still wondrous (if not also awesome and terrifying) and fundamentally beyond our human intervening powers, as an unending source of human value, significance, and meaning—a source to which we characteristically and recurrently return to refresh, recreate, and reeducate ourselves. Wild fish in wild streams experientially open humans on the vast, spatiotemporally deep universe that creatively, if amorally, engenders the streams, the fish, and the humans alike. This is Spinoza's or Darwin's *natura naturans*, "nature creative," divine substance or no, to which we all ultimately owe our very being. (Again, what may be amorally engendered is a moral issue if it is good [valuable] and vulnerable to human harm.)

Modern urban and technological civilization, and perhaps animal biotechnology in particular, remorselessly if unobtrusively threaten to overtake not only nature, but also the multivalenced natural symbols and naturally ingrained values by which we live and humanly sustain ourselves. Nature natural threatens to be eclipsed by nature humanly contrived. Nature evidenced and symbolized by wild and richly populated trout streams and salmon rivers threatens to become a farm pond of engineered and stocked fish frequented by humans with only practical, provincial, *polis* things in mind. This would be the further demise, begun in earnest by Descartes and others in the seventeenth century, of nature as "significant cosmos," as prehumanly and inherently valuable and alive.

This is the "impractical" cultural threat of animal biotechnology to traditional and recurrent symbolic orientations and to deeply felt ethical, aesthetic, and spiritual values originated in and provoked by the human experience of nature. It should be the business of philosophic cosmologies, in their defense of the full range of human and natural values, to counter such threats. Here is why the philosophically esoteric (or so it seems) practically matters so much. Philosophic worldviews can give limits or boundaries to our thinking, while recognizing the boundless goodness and significance of the dynamic togetherness of humans, animals, and nature.

Frog Pond Philosophy

What can I add to the ontologies and cosmologies of Spinoza and Darwin? Precious little. But perhaps I can throw in a comedic coda and a moral hope for the future.

Several years ago, after fishing an evening mayfly hatch for wild brook and brown trout, I sat by a northern Wisconsin pond, Brock Pond, sipping Old Grand Dad whiskey, smoking a cigarette, croaking along with the frogs—all good subversive behavior according to civilized, urban standards. Suddenly a philosophic lightening bolt flashed through my body. The universe burst forth into sound with the croak of a frog. Before there had been a vast, meaningless silence of whirling forces. Now there was sound, a sounder, and an audience, appreciative of the character and quality of the sound, no doubt laced with a subtle meaning and significance. Before a soundless universe; now a soundfull universe, a cosmological frog leap forward.

No doubt the details are wrong. Probably there were pre-frog murmurings to be overheard. But the basic philosophic insight of the cosmological emergence of sound (as well as life) seems right, and Brock Pond ever since has been for me sacred time and space for philosophic and moral reflection.

Lately, I have been thinking about our world and its evolutionary life as one vast, temporally deep frog pond, serving as the wider natural context of our humanly cultural adventures. I think back to the sixteenth and seventeenth centuries and the modern ontological and scientific revolution of Copernicus, Descartes, and company—a job left only half-finished. Perhaps Copernicus et al. moved the sun and earth away from the center of the universe, but thanks to personal gods, universal reason, language, other cultural talents, or whatever, humans still consider themselves at the center of all things significant and meaningful, right in the middle of the frog pond. Despite Darwin, little has changed. We "central ones" take all the natural resources that we need for our 6+ billion human selves, leaving what is left over for the rest of the pond's creatures, present and future. We industriously pursue our economies and technologies, spewing our wastes into the pond, recyclable or no. Few seem to care or even to know what we are doing, and all of us, like it or not, live more or less in the middle of the pond.

A few individuals, Aldo Leopold among others, have digested the Darwinian message and are morally and practically anxious to move us humans off-center, to find a seemly and right lily pad on which to live and croak, morally moved by a central concern for the overall and indefinite well-being of the frog pond as a whole. This is only a frog pond version of Leopold's land ethic, but neither have penetrated our urban centers and minds, where so much fateful human action and decision making take place.

Let me take one more philosophic turn around the pond in hope that it might somehow help. Prompted by the notions of orchestral causation and emergent individuals, I have begun to think of genomes, human or other, as scores of music, say a symphonic poem or Mozart's *Requiem*. Imagine that score in the hands respectively of elementary school, high school, college, and professional orchestras, soloists, and choruses. Given the differing contexts, we would expect different phenotypical results to emerge, more or less artfully expressed. So it is with uniquely diverse genomes in their uniquely different contexts.

But is not this realm of life, the frog pond, itself? There are billions upon billions of organisms with their innumerable genomes, a vast realm of historical interaction provoking the expressions of scores of natural music, some bursting into sound, some taking other naturally artful forms. What an incredible realm of historically engendered existence! So much value ongoingly bursting forth into being.

But then there is this problem at the center of the frog pond, that small section of the natural orchestra that refuses artfully and harmoniously to blend with the others, risking discordant cacophony in following its own tune. In fact, by its (our) misadventures, it is destroying or degrading the genomic scores of other individuals and species of life. (We are told we are amid a "human extinction" event.) The grand symphony of life and its future is being seriously marred and degraded. If we humans do not tune in, the pond literally might become frogless, humanless, soundless. Such a lapse into disvalue and lifelessness is a potential ontological and cosmological evil that vastly overshadows the pleasures, pains, life plans, and deaths of individual organisms, human or other. The wonderful and complex interplay of the bounded, boundaries, and the boundless threatens to become unraveled, leaving insignificant life or animate being, if any at all. Relatively or absolutely, there might be only the soundless, the valueless, the boundless. We would then need to replace the Benedictus of Mozart's *Requiem* with a contrite Ignoramus.

Urban philosophers and ethicists of mainstream utilitarian or deontological persuasions will be of little help here. They remain at the center of the pond, all too humanly bound and concerned. We need more measured reflections. We will need bullfrog philosophers who somehow can forcibly express the pond's natural music and the complexly intertwined and interactive symphonies that need to be protected and reinforced. Perhaps it will only be by such philosophic and moral music, or other value expression, that we will be able to penetrate and awaken urban ears, including our own. How else are we going to save nature's protean, but vulnerable sacred time and space, our earthly frog pond? This is urgent business which knows no bounds.

Appendix: Ethics and the Philosophic Cosmologies of Spinoza and Darwin

We have been arguing that a Spinozistic or Darwinian type of cosmology, with their fundamental conceptions of conative modal existence and organisms as emergents, which intricately tie together the basic themes of organic individuality and world relatedness, has decided philosophic advantages over a Cartesian, multiple substance metaphysics. The gain is on ontological and epistemological grounds and in lending a rational intelligibility to pervasive features of our experience of ourselves and the world. But what relation does such a cosmology have to ethics and, in particular, ethical responsibilities to the human, animal, and natural future?

The question is complex, but a first point is to understand what modern philosophic cosmology is and what it is not. Descartes and Spinoza notwithstanding, it is *not* the attempt logically to prove the existence and the nature of the world according to some mathematically inspired, deductive mode of reasoning. Philosophy is not geometric proof from rationally unassailable or self-evident first principles or premises. Rather philosophic cosmology is the speculative attempt to elucidate and further penetrate pre-philosophic human experience in a systematic way: to call to emphatic attention and to lend rational intelligibility and meaning to the fundamental features of the experience that we humans recurrently live through. (This is a post-Kantian or Whiteheadian conception of philosophic cosmology.)

The relation of philosophic cosmology to ethics similarly is not one of rational, logical deduction, with ethical theorems derived from ontological premises. The ethical "ought" is not *logically* derived from the ontological "is." Concrete ethical obligations and responsibilities aboriginally emerge from and within real worldly situations and not from rational, conceptual arguments. Yet there is a fundamental and substantive relation between what we conceive (ontologically) "to be" and what we consider ethically "ought to be" that can be spelled out by philosophic cosmology. This holds only if "what is" is primarily experienced and subsequently conceived as harboring its own values or goodness and, further, is understood to be vulnerable to (freely determined) human individual or collective action. If such is the case, ethical theory and action ought to cohere with our experiences and rational conceptions of worldly order, character, and goodness. It is the business of philosophy to spell out the possibilities of this coherence and to incorporate ethical experience and perceived responsibilities into its overall cosmological vision or worldview.

In short, powerful and persuasive philosophic perspectives focus our attention on and further elucidate concrete ethical obligations that hold (perhaps implicitly) within our worldly existence *prior* to rational and systematic reflection. Philosophy does not logically prove ethical obligations or oughts, but brings them more emphatically and sharply into focus. It argues why the obligations are intelligible, reasonable, and important. This is how we should pragmatically test the ethical relevance and adequacy of a Spinozistic or Darwinian type of interpretation of animal and natural existence. In the text, we fasten up one particular and interesting example, animal biotechnology.

Yet in considering the ethics of animal (and wider) biotechnology, we must note that a first task of philosophic cosmology, Spinozistic, Darwinian, or other, is to interpret and make rationally intelligible the very possibility of our being moral (and immoral) beings. We take Spinoza as our primary example, though the argument holds analogously for Darwin. With Spinoza, the interpretation rests on the notion of conative individuality, as lived out on a complex, human level. As with all organisms, we primordially endeavor to preserve in our own being, as necessarily implicated in the wider, dynamic world. Unlike other organisms, we become acutely and consciously aware of our own conatus and the conative implication of all animate entities in one world. Conative individuality, and all that is importantly caught up in conative endeavors, primordially strike us as major moral or ethical issues. (The title of Spinoza's fundamental philosophic work is *Ethics*.) Conative life necessarily concerns harms and benefits, goods and bads with respect to individuals and communities. "By nature and culture" this experience develops into a twin and interlaced concern for self and others: for the valuable or good whole, which includes naturally assertive individual selves. To be mutually concerned with self and other is the stuff, the warp and the woof, of ethical life. To be concerned disproportionately with oneself (or one's group) amidst a necessary and wider worldly relatedness is the "conative fall" into immorality or irresponsibility.

We need to explicate further the ethical import of the fundamental, twin themes of organic, conative individuality, and essential dynamic relatedness to the world. Conative individuality, in virtue of its vulnerability to harm and its realized natural goodness (characterized by individual existential effort conjoined with and realized through complex capacities, activities, and physical forms of organic being), is a primary locus of value and ethical concern. Natural or worldly communities, in virtue of being dynamically constituted by, essentially affecting, and themselves vulnerable to conative individuals, are an equally primary locus of value and ethical concern. Thanks to universal dynamic interdependence, conative individuals and worldly communities mutually engender, evolve, and require one another, as within the wider natural universe. (This is a primary meaning of internal relation or relatedness.) An ethical obligation or responsibility to uphold the ongoing, mutual, interdependent, and vulnerable goodness of conative individuality and worldly community sets the fundamental terms of the moral landscape.

Secondly, conative individuals and the concrete communities that they foster are characterized by varying levels of richness and complexity with respect to the realization of natural goodness or value. In the primordial obligation to uphold natural goodness and prevent significant and irreparable harm to the animate and conative realm, there is a certain differentiation and relative weighting of substantive ethical responsibilities.

With this second feature of the moral landscape we should note the relation of concrete values and natural goodness to worldly becoming and the rigor of animate existence: its "evil" or amoral harshness. The naturally good (the definite forms and active powers of individual and communal animate existence) and the naturally evil (the infliction or suffering of pain, death, or destruction of realized order) cannot be disjoined. Both symbiosis (cooperation and coordination) and competition (strife) are at the heart of organisms' metabolic mode of existence, natural evolutionary and ecological processes, and the interdependence of all things natural. It is not for naught that Heraclitus long ago declared that (ontologically) "War is father of all." This necessary connection of the darker side of existence and nature's goodness is an essential and inescapable given for ethical reflection and decision making.

The third feature of the moral landscape follows from the first and second. The precariousness and immediate demands of conative life require that one primarily take care of one's own: oneself, one's family, one's local community (human and natural), nation, species, etc. This is a pre-moral or existential, as well as moral, injunction. But the fact that all animate beings instance varying levels of natural goodness and are dynamically and internally related means that the primary, more immediate conative concerns ought ethically to be tempered by wider, less proximate obligations, for example, to spatially or temporally distant individuals or communities, human and natural.

So far we have offered no hard and fast, brightline principles to decide rationally particular cases of conflicting ethical obligations in a clear-cut and straightforward fashion. Nor may we legitimately expect to arrive at such final rational principles and avoid reliance on fallible moral wisdom and judgment. The real world is too bewilderingly complex and its realized values too many and diverse to be readily subdued by ethical rationality or reasoning. We may ever need to learn an art of moral ecology: to allow in particular and concrete contexts conflicting obligations or responsibilities to play themselves against one another, mutually determining each other's relative weight or strength, until we arrive at a morally satisfactory (if imperfect) judgment, "all things considered." Yet we can discern substantive and important general ethical directions or guides, expressible in ethical injunctions. Uphold the ongoing and variously rich animate realm as a whole. Prevent or address significant harms before hazarding improvements on the presently realized good.

(Though vulnerable to human harm, the conative realm amorally takes care of itself prior to human, "ameliorating" interventions.) In particular, beware the ethically harmful systemic effects of particular and necessarily limited corporate human purposes and actions. Give primary ethical attention to the integrity or intactness of conative individuals and worldly communities in their dynamic and temporal interconnections.

Such ethical imperatives and the philosophic cosmologies that underscore them concentrate our ethical attention and propose what we ethically ought to take into account: again, the fundamental, though varying significance of organic individuals (human and other) and worldly communities human, animal, and ecosystemic, specifically their worldly needs, integrity, and resilience into the indefinite worldly future. Again, all this holds if earthly life is deeply, characteristically, ultimately valued as good.

References

Rene Descartes. 1991. *The Philosophic Writings of Descartes*. Volumes I and II.. Translated by John Collingham, Robert Stoothoff, and Durgald Murdoch. New York: Cambridge University Press.

Strachan Donnelley. 1993. "The Ethical Challenges of Animal Biotechnology." *Livestock Production Science* 36 (July):1.

———. 2000. "Nature, Freedom, Responsibility: Ernst Mayr and Isaiah Berlin." *Social Research* 67 (Winter): 4.

———. 2001. "Philosophy, Evolutionary Biology, and Ethics: Ernst Mayr and Hans Jonas." *Graduate Faculty Philosophy Journal* (New School University), 23:1.

———. 2002. "Leopold's Darwin: Climbing Mountains, Developing Land." In *The Good in Nature and Humanity*. Stephen Kellert and Timothy Franham, eds. Washington: Island Press.

Strachan Donnelley, Charles R. McCarthy, and Rivers Singleton, Jr. 1994. "The Brave New World of Animal Biotechnology." Special Supplement, *Hastings Center Report* 24 (January–February):1.

Hans Jonas. 1987. *The Imperative of Responsibility*. Chicago: University of Chicago Press.

———. 1982. *The Phenomenon of Life*. Chicago: University of Chicago Press.

———. 1980. *Philosophical Essays*. Chicago: University of Chicago Press.

Ernst Mayr. 1991. *One Long Argument*. Cambridge; Harvard University Press..

———. 1997. *This Is Biology*. Cambridge: Harvard University Press.

Benedict de Spinoza. 1951. *Ethics*. Translated by R. H. M. Elwes. New York: Dover Press.

Chapter 4

Cutting Nature at the Seams: Beyond Species Boundaries in a World of Diversity

Jon Jensen

Few concepts play a more central role in contemporary scientific thought than the species. Species are considered the foundation of both taxonomy and evolutionary theory in biology as well as the most significant unit of life within common language. Species play a similarly significant role in the political and conservation arenas. The Endangered Species Act (ESA) is a cornerstone of American environmental policy, and species extinction is the prime fear driving efforts to halt destruction of the rainforest and many of the other bogeymen of the environmental movement. This combined political and scientific significance makes the species one of the centerpieces of human thought as we begin the twenty-first century. While I have generally shared this view of species, lately I have begun to wonder if this role for species is truly justified. Should we be placing so much scientific and political emphasis on species boundaries?

I first began to question the dominance of species in thinking about wolf reintroduction. The wolf is a vitally important symbol of the challenges faced by efforts to protect native biodiversity in a modern and pluralistic world. Wolves are loved and revered by many people— for being wild and beyond human control—but also hated and feared by others for many of the same reasons. The reintroduction of wolves into Yellowstone and Idaho a few years ago brought all of these emotions to the surface, illustrating the power of this species and all that it symbolizes. When a proposal surfaced to reintroduce wolves into New York's Adirondack Park the same emotions boiled over in a different setting. The northeastern wolf case also presented an interesting example of another, more scientific dispute regarding the boundaries between species and the role that species play in our efforts to conserve native biological diversity.

Wolves were once abundant in the Northeast, as they were throughout most of the United States, but were extirpated from New England by humans roughly 100 years ago. Although we know that wolves once roamed this region, it isn't clear exactly which species of wolf was here. It has long been assumed that the gray wolf, *Canis lupus*, was the top predator in New England, but recent studies have raised serious doubts about this conclusion. The last known wolf killed in the Adirondacks shows mtDNA similar to that from the species *Canis rufus*, the red wolf, while other studies indicate links to the wolves of Canada's Algonquin Provincial Park (Wilson et al. 1999). Were the northeastern wolves red wolves? Were both red wolves and gray wolves here? Or, perhaps, was the eastern forest wolf a different species entirely, one that is now gone from the United States?

The situation becomes even more complicated with the realization that coyotes, *Canis latrans*, must also be included in this mix. Though coyotes were not present in this area prior to European settlement, they have emigrated to the Northeast in large numbers in the last fifty years and have thrived in the void left by the extirpated wolves. The coyote has thus largely moved into the ecological niche that was once occupied by wolves in the Northeast. Beyond this ecological reality is the confusing factor that the species boundaries between wolves and coyotes are not always clear. Eastern coyotes, unlike their smaller Western cousins, are similar in size to some wolves, especially the red wolves. Thus, the possibility exists—in the past, present, and future—for interbreeding between wolves and coyotes, with the potential to create hybrids between the two species (Roy et al. 1994).

While these uncertain species boundaries are not some major scientific dilemma, they do have ramifications for efforts to reintroduce wolves into the Northeast. If we are not sure which species of wolves are native to the Northeast, how can we know which wolves to reintroduce? And if we are worried about interbreeding with coyotes, isn't that a powerful argument against reintroduction because of the fear of hybridization? Clearly the main opposition to wolf reintroduction is political, not biological, but it would be a mistake to see these two as completely separable. Biological disagreements over reintroduction have fueled political opposition and resolving the biological dispute is necessary, if not sufficient, for overcoming political obstacles. The reality is that until we are able to sort out the issues of taxonomy, ecology, and evolution inherent in concerns regarding wolf species in the Northeast, efforts at reintroduction are at a standstill.

The case of wolves in the Northeast raises a number of broader questions about species, conservation, and boundaries. For example, what exactly is a species, and how do we determine the boundaries between different species and between species and taxonomic units such as subspecies and genera? Why are species considered a unique level of diversity in both science and policy? Is our current understanding of species boundaries consonant with a fully evolutionary perspective? More important, is a species-based approach sufficient for conserving and restoring the full range of biological diversity and the evolutionary and ecological processes on which it depends? While our system of taxonomic boundaries made perfect sense within the pre-evolutionary mindset of Aristotle and Linnaeus, it is less clear that it should be maintained given our current understanding of evolution.

What Is a Species?

I have heard it said that regardless of what question you ask a Vermont gardener, the answer is always the same: "It depends." "Can I grow asparagus in Vermont?" "It depends." "When should I plant my rutabagas?" "It depends"—on what part of the state, on the weather, on the elevation, on the year, on your soil, on.... It depends.

The same might be said for the question of what distinguishes an organism as a member of a particular species: it depends. It depends on the group in question, on what species concept is being used, on what you are studying or care about, on the historical trends within the field, and above all on whom you ask. One searches in vain for truly universal rules or trends within the classification of the vast array of living organisms on the planet. While this pluralism rarely seems to trouble most biologists, it is a bit disconcerting to some people who worry that it undermines the notion that science is objective, describing reality rather than constructing it. If species boundaries are not based on universal criteria, they worry, doesn't that mean that they are arbitrary human constructions?

The notion of a species has not always seemed so variable. While the word "species" originated in the seventeenth century, the idea of identifying "kinds" of plants and animals and drawing boundaries between different groups of organisms dates back nearly as far as written human history. Until the nineteenth century, species were generally seen as discrete and fixed units, what philosophers call "natural kinds." This view was often religious in origin, associated with the belief that God created all living things, each in its own kind, clearly separate from the rest.

Aristotle is probably the best-known early taxonomist, naming, sorting, and classifying living things. Based on his observations, he created groupings of similar creatures based on appearance, which he hierarchically arranged in his "ladder of life," the first formal taxonomy. All living things, Aristotle believed, belonged to some natural kind with an essence that defined what it meant to be a turtle, an apple tree, or a person. Central to the Aristotelian system were the twin ideas of essentialism and the fixed nature of species. A species was defined by an essence or ideal type, which separated the individuals in this species from members of other species. This essence of a species was unchanging for Aristotle. That species are fixed entities with defining essences is an idea that died hard within science. Species essentialism had an even stronger hold within a largely religious-minded general public where the idea that "the world comes to us prepackaged into units" (Kitcher 2001) seems to be alive and well still today.

A more modern, scientific perspective on species began to develop in the nineteenth century. Darwin's work spurred on the vigorous debate regarding the proper understanding of species, since the idea that species are fixed was clearly incompatible with evolution through natural selection. This debate continues, as lively as ever, with no solution to the "species problem" on the immediate horizon.[1] In fact, the number of species concepts keeps expanding rather than contracting toward consensus. There are more than twenty different species concepts actively defended in the literature today in over a hundred books and articles (Ereshefsky 1992a). Moreover, many of these different species concepts produce different and incompatible taxonomies. Thus, the debate over species concepts is more than purely academic, affecting how we classify, study, and understand biological diversity.

While the numbers are large, species concepts can be grouped into broad categories for basic understanding. Perhaps the most widely accepted approach is the Biological Species Concept (BSC), which is based on the idea of interbreeding as the key differentiating criteria between species. Ernst Mayr first advocated this concept in his 1942 book *Systematics and the Origin of Species* and in his many articles and books that followed. The BSC attempts to define species in such a way that it is not our perceptions of similarity and difference but the discriminations of the organisms themselves that are the key factor. Species, according to Mayr (1970), are "groups of interbreeding natural populations that are reproductively isolated from other such groups." On this view, if two organisms are able to interbreed in the wild and produce viable offspring, they are the same species. If they do not interbreed, they are separate species.

A second approach, phylogenetics, focuses on descent from a common ancestor, or monophyly, as the determining factor for species. Each lineage descended from a common ancestor is called a monophyletic taxa, and species are generally considered the basal monophyletic taxa. Phylogeneticists often divide organisms based on morphological differences, though the intent is to use multiple characters for analyzing distinctiveness. Thus, species boundaries are often determined based on similarity of appearance or structure. Two individuals are assumed to be members of the same species if they share the same appearance and morphological structure since this is generally assumed to be evidence of descent from a common ancestor. In a sense, what contemporary phylogeneticists are doing is using similar features as a proxy for determining shared evolutionary lineage. Thus, if two toads share the same basic physiological features, then they must have evolved from the same common ancestor, whereas differing features indicate differing lineages.[2]

A final, though less popular, approach to defining the species concept looks to ecology for its definition. An ecological species concept (Ehrlich and Raven 1969, Van Valen 1976) focuses on environmental factors, not reproduction or similarity, as the key to the boundaries between species. According to Van Valen (1976), a species "is a lineage . . . which occupies an adaptive zone minimally different from that of any other lineages outside its range" (235). Species are thus defined by ecological niches and the unique set of selection forces within that niche or adaptive zone. What keeps species separate is the power of these selection forces. Darwin's finches provide an excellent example of this approach. The different species of finches he studied were separable based on the different ecological conditions that had driven their evolution. While such ecological conditions often result in morphological differences—beak shape and size, for example—the defining feature is the adaptive zone occupied by that species.

Though these species concepts are radically different in some ways, they also share much in common. All mainstream species concepts, for example, share the view that species are lineages. Thus, whether we draw species boundaries based on interbreeding, ecology, or other factors, we still have descent and evolution as shared assumptions in a common understanding of species as lineages. Advances in genetic testing also indicate a root similarity among the various species concepts. Since species are lineages, similarity in genetic makeup between organisms is a good indicator of species boundaries. Genetic analysis is used by advocates of differing species concepts though it may play different roles in separating species taxa. Given these basic similarities, convergence between the different species concepts is clearly the norm in determining species boundaries. It is not, however, universal. And while disagreements about how to draw boundaries between species may be rare, they are real and, at least in some cases, very significant.

The eastern forest wolf provides an interesting case of difficulties that exist in separating species as well as the enormous stakes riding upon such boundaries. While the proper wolf taxonomy may seem clear to most people, recent studies are leading researchers toward different theories about how to divide up the canids of North America. The standard view is that there are two species of wolves in North America—the gray wolf, *Canis lupus*, and the red wolf, *Canis rufus*. Prior to human extirpation of wolves throughout the majority of their native range, gray wolves were common in most of the northern half of the continent while red wolves roamed throughout what is now the southeastern United States. Gray wolves have been divided into as many as twenty-two different subspecies at times, but now there are five accepted subspecies of *Canis lupus*: *arctos* in the Arctic northeast, *occidentalis* in the Pacific Northwest, *baileyi* in the desert south, *nubilis* in the central continent, and *lycaon* in the Northeast. On this theory, the eastern wolf of New England would have been a gray wolf of the *lycaon* subspecies, more formally *Canis lupus lycaon* (Brewster and Fritts 1994).

Other biologists (Wilson et al. 1999) have offered an alternative theory of wolf species boundaries based on their analysis of mtDNA in wolves and coyotes. This group proposes a new species of wolf, *Canis lycaon*, separate from the gray wolf and including but not limited to the *lycaon* subspecies. They argue that the red wolves of the South and the smaller wolves of eastern Canada are sufficiently similar to be classified as one species. If you include the extirpated wolves of New England—which this group claims to be the same species—we have a continuous geographic distribution of this species from eastern Canada through the eastern United States.

Yet a third drawing of wolf species boundaries relies heavily on the assumption that wolves and coyotes have hybridized extensively and argues that all North American wolves are really one species. Robert Wayne and his colleagues (1995) suggest that both red wolves and the troublesome *lycaon* subspecies are not true species but rather hybrids. This theory assumes that the smaller species are a result of interbreeding, not true specific differences. On this view, the wolves of the Northeast are not true wolves at all. Rather, they are not significantly different from the coyotes that are common in the area already. Thus, there is no need for reintroduction of wolves into the Adirondacks or anywhere else in the Northeast.

Which taxonomy is the correct one? What species, subspecies, or hybrid were the wolves of the Northeast and hence the proper animals to be reintroduced? It may be tempting just to shrug off these disagreements over wolf taxonomy as one more in the long list of scientific disputes that are a natural and healthy part of the workings of the scientific method. After all, science is supposed to sort competing hypotheses, taking us closer to the truth in small but true steps. Perhaps, but I cannot help wondering whether there is *one* right answer, to wonder, that is, if science is able to solve this disagreement in a definitive way. In this case quite a lot depends on which hypothesis wins out, on which drawing of species boundaries prevails with North American canids. Not only the controversy over reintroduction, but also questions of how to manage habitat depend on the resolution of this taxonomic dispute. The reality is that different species concepts, different readings of the evidence, and different assumptions yield different species boundaries. These boundaries in turn result in different answers to the questions that must be resolved along the road to restoring the native biodiversity and healthy ecosystems of the Northeast. These questions about species cannot be ignored since they influence questions of land management, shaping the way that we use—or do not use—public and private lands.

Species Pluralism

When faced with a range of different species boundaries, as we are in the case of wolves, we have two options. Either we can argue that one delineation of species taxa is true and complete, while all of the others are mistaken, or we can opt for pluralism as the best solution to this aspect of the "species problem." Instead of trying to settle on the one true way of determining species boundaries, the pluralist answers "it depends" to the question of the proper understanding of a species and thus the proper species boundaries. While species pluralism has been widely criticized by philosophers of science, its acceptance by biologists seems difficult to deny. Even an introductory college biology textbook claims "no classification system should be viewed as *the* system," noting that "as long as observations continue to be made, different people will interpret relationships among organisms in different ways"(Starr and Taggart 1989).

In broadest terms, species pluralism is the view that there are multiple, valid species concepts and thus multiple acceptable criteria for determining species boundaries. While the implications of this view seem troubling to some, the view itself is relatively straightforward. Both interbreeding and ecological species concepts are acceptable, for example. Unfortunately, however, the plurality of types of species pluralism quickly muddies the waters. Some types of species pluralism are hardly distinguishable from the monistic claim that there is one true taxonomy of the world. Michael Ruse, for example, argues that the various species concepts all coincide, picking out exactly the same organisms even though they are using different criteria. "There are different ways of breaking organisms into groups, and they coincide! The genetic species is the morphological species is the reproductively isolated species is the group with common ancestors" (Ruse 1987). While a very tidy picture, Ruse's claims do not fit so well with observed reality. Groups of organisms that interbreed are often not the same groups with the most genetic similarity, the ones possessing monophyly, or the ones that form ecological units. Wolves provide one example of this lack of consilience, but there are countless others.[3]

Another form of pluralism recognizes that the different criteria for sorting organisms into species do not always coincide, but argues that one factor is "most important" for each group of organisms. Thus, reproductive isolation may sort mammals into species, while ecological forces determine certain plant species, and morphological differences are the key for protists. As Mishler and Donoghue (1982) argue, "a universal criterion for delimiting fundamental, cohesive evolutionary units does not exist" (495). While all species are the result of evolutionary lineages and there is a single taxonomy of the world, this view is pluralistic because it recognizes that different factors may be most important in the evolution of different groups.

Philip Kitcher's pluralism is similar though more pragmatic in its focus. According to Kitcher (2001), "We divide things into kinds to suit our purposes" (49). An ecologist, for example, primarily concerned with energy transfers, might divide organisms one way while the biochemist will take a different tack. While some groupings are clearly artificial, it is not the case that there is one correct system of classification. In discussing how different groups of medical researchers use differing criteria to classify the various organisms responsible for human disease, Kitcher is incredulous at the notion that only one species concept is correct.

> It would be absurd pedantry to insist that a single way of classifying organisms must take precedence and that one of the taxonomic schemes is "unnatural." For the purposes of the classification both are obvious and well motivated: researchers want to divide up the organisms in ways that help combat human disease. Once again, the partitioning of nature accords with our interests and, in a less obvious way, with our capacities. (49)

The most radical forms of pluralism suggest eliminating the very concept of a species as obsolete and misleading. Marc Ereshefsky (1992b), for example, defends what he calls "eliminative pluralism" since it would "eliminate the term species, and replace it with a plurality of more accurate terms." According to Ereshefsky, systematics provides not only multiple species concepts, but also "a plurality of equally legitimate, though incompatible, taxonomies of the organic world." Ereshefsky argues that the various "forces of evolution segment the tree of life into a plurality of incompatible taxonomies: one taxonomy consisting of interbreeding units, another consisting of ecological units, and a third consisting of monophyletic taxa." None of these taxonomies is preferable, however, and thus species concepts overlap with some individuals belonging to more than one species. In responding to the obvious objection that this will result in confusion in communication between scientists, policymakers, and even the general public, Ereshefsky suggests a radical solution: abandon the concept and the word "species" altogether.

> Instead of referring to basal lineages as "species," biologists should categorize those lineages by the criteria used to segment them: interbreeding units, monophyletic units, and ecological units. The term "species" is

superfluous beyond the reference to a segmentation criterion; and when the term is used alone, it leads to confusion. The term "species" has outlived its usefulness, and should be replaced by terms that more accurately describe the different types of lineages that biologists refer to as "species." (Ereshefsky 1992b)

Instead of species we should talk about "ecospecies," "biospecies," and so on, depending on the criterion used to demarcate the species boundaries. Thus, Ereshefsky suggests, we should be pluralistic in the words we use to describe the species rank as well as the species concepts and resulting taxonomies. While I find Ereshefsky's proposal to be too extreme, it raises a very important question: why do we consider species and other taxonomic ranks to be nearly sacred elements of the biological world? Answering this question requires a brief detour into a little history.

The Legacy of Linnaeus

While multiple species concepts muddy the waters today, things were once much clearer. The essentialist species concept, the idea that species are fixed natural kinds, dominated biological classification into the nineteenth century. This understanding of species formed the foundation for the Linnaean system of classification, which has dominated biology for over 200 years, showing surprising resiliency. Both the binomial system of nomenclature—Latin genus and species names for all creatures—as well as the ranked hierarchy of Kingdom, Phylum, Class, Order, Family, Genus, Species owe their existence to Linnaeus. The widespread recognition of the binomials for species such as wolves—*Canis lupus,* for example—shows how deeply embedded the Linnaean binomials are in scientific and even popular language. Though Linnaeus's classification was based on morphological similarities, not evolutionary relationships, the Linnaean hierarchy has survived more or less intact into the era of evolutionary biology. But should it? Are there reasons besides habit and inertia that justify the Linnaean ranked system? Perhaps, but even beyond issues related to species concepts, problems exist for the entire Linnaean ranked hierarchy. A few examples should serve to illustrate these difficulties.

First, as with species, there is very little uniformity in the way that the ranks of the Linnaean system are applied. Even a basic zoology textbook acknowledges this lack of consistency.

> Above the species level, there are no precise definitions of what constitutes a particular taxon. Disagreements as to whether two species should be grouped into the same taxon, or different taxa are common. Some biologists

(splitters) prefer to have taxonomic categories with a few, closely related species included. Others (lumpers) prefer categories with more species. (Miller and Harley 1992)

While this variability in application may not be a problem on its own, it underscores the fact that the system is ultimately awkward and is based on human preferences just as much as evolutionary lineages.

A further problem results from the Linnaean insistence on a set number of ranks within each kingdom, regardless of the range of diversity. Thus, both the animal and plant kingdoms must be fit into the same fixed ranks in spite of the enormous differences in the number and types of organisms. The only way that different taxa can be shoehorned into the same ranks is through a combination of creativity and arbitrariness that belies the supposed "orderliness" of the Linnaean system, its primary justification. The creativity comes in the profligate use of subranks by many taxonomists in order to fit the organisms of wide ranging evolutionary lineages into the set seven ranks. In some cases the amount of diversity does not warrant all seven ranks: bison, for example, are the only species in the genus. In other cases (Pacific salmon, as I will discuss later), there are more subranks than primary ranks in order to account for the range of diversity. While the awkwardness and ad-hocness inherent in this system is problematic, the true difficulty arises from the inevitable result that the ranks have different meanings for different taxa. Thus, the rank of family in classifying orchids (*orchidacae*) is in no way equivalent to the rank of family for weasels (*mustelidae*), and the disconnect only grows greater when we move to the insects. It's not simply that there are more species in one family than the other, but that the family rank represents very different points in the evolutionary lineages of these two groups of organisms. As Brent Mishler (1999) has argued, this non-equivalence of rank makes the Linnaean ranking at best "a meaningless formality" and at worst it can lead to bad science.

Despite its historical importance and resilience over time, the whole Linnaean system is very problematic simply because it's misleading. It assumes comparative order where none exists. Through evolutionary processes, lineages are constantly dividing, but there is no reason to think that these lineages are directly comparable at different points in the evolutionary trajectory. It is only the unceasing human attempt to impose order onto nature's "messiness" that motivates and perpetuates the Linnaean system. Thus, if our current understanding of species rank rests upon its role within the Linnaean system, this is a very tenuous foundation.

The reason, I suspect, for the longevity of both the Linnaean system and our understanding of species is neither their accuracy nor objectivity, but simply inertia. This is what all of us learned in school, and it's familiar.

Besides, we need some classification system and we lack a viable alternative. This is a mistaken impression, however, as vocal opponents of the Linnaean taxonomy have called for a "rank free" system of ordering biodiversity (Mishler 1999). The problem with the Linnaean system, according to these critics, is neither the ordering nor the hierarchy; hierarchies are an essential element of a classificatory scheme. The problem comes from the fixed ranks applied to the hierarchy within the Linnaean system. Instead of ranks, we should simply identify the varying levels of the evolutionary lineage of any given group of organisms. This could be done through assigning a proper name to each level of the hierarchy for the group, but not ranking the various levels. For most organisms we could refer to them simply by the current species name—sapiens for humans, for example—though at times we would need to refer to the full names, which would be reverse orderings of Linnaean categories plus or minus depending upon the group in question. Thus, the full name for humans would be Sapiens Homo Homidae Primate Mammalia Vertebrata Metacoa Eucaryota Life, under one rankless scheme (Mishler 1999).

While the differences between this scheme and the traditional Linnaean ordering may seem slight, they are significant and may well be worth the fuss required to change what has been the status quo for over two centuries. Such a change could help us with the ongoing process of shaking free from our deeply ingrained habits of thought, which are contradictory to evolutionary biology. Despite efforts to the contrary, Linnaean classification assumes fixed types and trajectories for all creatures that mirror those of all other organisms. As with the species rank, other taxonomic ranks are ultimately human judgments about boundaries within a world of diversity. It is not that there are no breaks or divisions within nature; clearly there are. Rather, the problem is the claim that we can identify objective points on the diverging evolutionary lineages of organisms that are more important or more significant than others. Whether we call these families, genera, species, or evolutionary units, there is no difference in importance. All levels of division within a hierarchy of evolutionary lineages are equally important, a fact that is only obscured by the pre-evolutionary ranking of the Linnaean scheme of classification. Thus, not only is the Linnaean system not a justification for species boundaries, but such a system needs to be reformed to more truly reflect the reality of an evolutionary worldview.

Once we abandon the Linnaean system, however, it is unclear why we should treat species differently from other taxonomic units. I suspect that the strongest argument for our belief in fixed species boundaries is neither taxonomic nor evolutionary. In fact, it's not scientific at all. Probably the strongest argument for focusing on species is pragmatic. We are accustomed to talking about, thinking about, and studying species. This is what we have always done, and it's what we know how to do. Focusing on species is familiar not only in science but also in our efforts at conserving biodiversity. A shift away from species would shake things up in a way that is unpredictable and thus potentially dangerous, especially in regards to endangered species. This practical argument is quite strong, in my opinion, and not to be taken lightly. Neither, however, should it be accepted uncritically.

Conservation and Fuzzy Boundaries

If my previous argument is correct, then we should move beyond traditional taxonomic ranks and boundaries in thinking about biodiversity. It remains, however, to consider the impact of this on conservation where species are typically the central unit of biodiversity. If the concept of a species is contested terrain and the taxonomic classification system is flawed, where does this leave the Endangered Species Act? Doesn't my argument undermine our premier piece of conservation legislation?

I don't think so, though it should help us to rethink conservation both within and beyond the Endangered Species Act (ESA). First, we must remember that all of these conflicts over species predate the ESA and have raged throughout the thirty years since it was enacted. Pluralism regarding species has been the reality within biology for decades and will remain so. The result has been a case-by-case approach to species proposed for listing, and there is no reason to think that this will (or should) change. We must look to the best available science to draw the boundaries and make determinations about the status and prospects of species. My arguments in no way entail (or even imply) that we should abandon or weaken the Endangered Species Act. On the contrary, an approach to conservation that moves beyond species is fully in keeping with both the letter and the legislative intent of the ESA. In addition to protecting the individual members of all threatened and endangered species, the law mandates adequate protection for the "critical habitat" that these creatures need to survive. Moreover, the ESA allows for the listing of subspecies and even "distinct population segments" (DPSs) if these are shown to be evolutionarily significant. This last provision, though not used often enough, provides a helpful example of how the scope of the ESA should be extended beyond species.

The fact that the ESA allows for conservation beyond species only serves to underline the fact that a species-only approach to conservation is insufficient. Even with all species protected, thousands (if not millions) of subspecies, races, varieties, and distinct populations will go extinct, and countless evolutionary possibilities will be lost. While the species approach embodied in the ESA has definitely worked—without it hundreds or thousands of species would have been lost in the last thirty years—it is pure folly to think that it is sufficient. Once we see that the species rank is not special, the need for conservation beyond species becomes clear.

Though we must maintain the integrity of the ESA and expand its reach, it is essential that we shift the focus of conservation beyond the species level. We need gradually to unseat the assumption that species are the only significant level of biological classification. Rather, we must look simultaneously both up and down the taxonomic hierarchy as we recognize the range of biodiversity and evolutionary processes. How do we make these shifts? First, if the ESA is going to reach its full and intended potential, much more needs to be done to list subspecies and distinct population segments as threatened and endangered. Beyond improvements in these current practices we should shift our focus toward evolutionary units as the basic units of consideration. Such a move should provide greater impetus to current efforts to gain ESA protection for taxonomic levels below species.

Second, we should go beyond the Endangered Species Act to enact legislation specifically targeted at other levels of biodiversity. The only way to protect the fullest possible range of natural diversity is to push the focus of our conservation efforts in two directions: both down to smaller taxonomic and evolutionary units and up to a more holistic, ecosystemic approach. As we push lower (toward the evolutionary unit) and higher (toward an ecological focus), these two approaches should unite in a circle broadening and deepening our efforts to protect the full array of natural biological diversity. A further discussion of both the upward and downward movement should help to show how this could work.

The best example of downward movement in the ESA, toward populations, is the focus on "evolutionary units" (EUs). The shift toward EUs was recommended by the committee of the National Research Council (NRC) commissioned to study *Science and the Endangered Species Act* (1995). An evolutionary unit, according to the NRC, is "a group of organisms that represents a segment of biological diversity that shares evolutionary lineage and contains the potential for a unique evolutionary future"(NRC 1995, 57). Central to the concept of an EU is "distinctiveness" since "a basic characteristic of an EU is that it is distinct from other EUs." The NRC identifies the following five characteristics that determine distinctiveness: (1) genetic isolation; (2) geographic isolation; (3) temporal isolation; (4) behavioral isolation; (5) reproductive isolation. These criteria do not, however, fit into a neat formula, but must be considered "holistically with no one level to be dispositive of the determination of a population's qualification as an EU."

The Evolutionarily Significant Unit or ESU is a closely related concept used by the National Marine Fisheries Service in its work to identify and preserve marine life. ESUs have been used most notably in the case of Pacific salmon. Each distinct run of the seven species of salmon—214 in all—has been identified as significant from an evolutionary perspective and thus deserving of protection. While the EU and ESU concepts do differ—EUs deemphasize reproductive isolation and place greater weight on evolutionary future—the differences are small and both concepts are likely to lead to very similar results for most groups of organisms. While species are often EUs, subspecies and populations are given equal or greater weight when they are distinct and thus evolutionarily important. This system helps to bring the Endangered Species Act more in line with an evolutionary worldview and the corresponding changes in taxonomy. Such a framework is not only consistent with but seems to better reflect the goals of the ESA.

Simultaneous with this consideration of units below the species level, we must also shift the focus upward toward a more holistic approach. A good practical step in this direction would be the adoption of a Native Ecosystems Act as proposed by Reed Noss and the Biodiversity Legal Foundation (Noss 1999). With its more inclusive and holistic perspective, a Native Ecosystems Act would "serve to protect and restore the entire spectrum of native plant and animal communities across the United States" (31). By focusing on ecosystems rather than species, this sort of approach has the potential to reap tremendous benefits in the protection of biodiversity. Even here, however, we cannot avoid boundaries and hierarchies. As Noss notes: "Conservation of ecosystems must begin with a comprehensive, hierarchical classification of ecosystems for the region of concern—in this case the United States" (ibid.). Such classifications have begun in many places already through the work of The Nature Conservancy and other organizations. This classification might in some respects serve as a model for taxonomic classification since it appears to contain no assumptions about a base or fundamental level of concern.

In addition to new legislation focused on protecting entire ecosystems, we also need a greater focus on protecting all genetic diversity. Conserving a species but losing ninety-five percent of the original genetic diversity within that species is not preserving biodiversity, and this is potentially a major problem with species-based approaches. An approach to conservation that moves beyond species should explore other possibilities such as a Genetic Diversity Preservation Act (GDPA). Such an act would provide a mechanism for addressing all losses of genetic diversity caused by human intervention. The GDPA could also provide help in preserving genetic diversity in agricultural crops and livestock where homogenization poses significant threats to the stability and health of our food supply.

While initially scary, abandoning the notion that species are special opens up new possibilities for biodiversity protection. The possible payoffs of these new opportunities—more focus on distinct population segments and evolutionary units within the ESA; a shift in emphasis toward considering whole ecosystems as well as all segments of genetic diversity—while difficult to achieve politically are well worth the risk and effort. A look at a couple cases—Pacific salmon and northeastern wolves—should help to show how conservation beyond species boundaries can and does work. We see in these cases how the Endangered Species Act is already going far beyond species and how these efforts may point the way to further advances.

Of Wolves and Salmon

Pacific salmon, which originally occupied every tributary of coastal rivers in the Northwest, are now extinct in forty percent of their historic range (Lichatowich 1999). Perhaps more important, nearly half of the remaining populations are in some danger of extinction. What is interesting about the case of Pacific salmon is that none of the seven species themselves are in danger of extinction. Rather it is the distinct populations over which so much concern is rightly centered. Here is an example of how shifting our focus away from species was necessary to address the threats to the organisms and a possible shift back to species is threatening conservation efforts.

There are seven distinct species of Pacific salmon: pink, coho, chinook, chum, sockeye, steelhead, and sea-run cutthroat. All of these species are members of the Genus *Oncorhynchus* in the family *salmonoidei*. In addition to the standard categories, the full Linnaean classification for Pacific salmon includes Infraclass, Division, Superorder, and Suborder. This provides a good example of the problem inherent in the process of trying to fit groups of organisms into the Linnaean system given advances in evolutionary biology and systematics. Within these seven species is tremendous genetic, evolutionary, and ecological variation. In fact, studies have identified 214 native, naturally spawning runs of Pacific salmon in the Northwest. Unfortunately, none of these runs is doing particularly well. A 1991 study, "Pacific Salmon at the Crossroads," breaks down the status of the various populations as follows: 101 "were at high risk of extinction"; 58 "at moderate risk"; 54 "were of special concern"; and 1 run had already been protected under the ESA (Lichatowich 1999, 204).

Is it acceptable if a run is lost to extinction so long as the species remains? The obvious answer from salmon biologists, with which the National Marine Fisheries Service (NMFS) concurred, was no. For this reason the separate runs were listed as evolutionarily significant units (ESUs); species or not, these runs are significant and must be protected. Starting with the first listing of the chinook in the Sacramento River in 1989, more and more of these ESUs have been individually protected under the Endangered Species Act. Without this approach—focusing on ESUs rather than simply species—it is difficult to imagine how much more genetic diversity would have been lost. Pacific salmon seem to provide a perfect example of how conservation beyond species can and does work within the framework of the ESA.

Unfortunately, this story is far from over, a happy ending is anything but certain, and an excessive focus on species may be part of the problem. While hatcheries have played a crucial role in supplementing stocks of wild salmon, their role has been controversial, and hatchery salmon may now threaten the survival of their wild cousins. The problem stems from a lawsuit charging that the NMFS was negligent when they failed to include hatchery fish in their assessment of the status of coho salmon. The argument is simple: since the hatchery fish interbreed with the wild salmon and contain similar genetic material they are the same species and thus must be counted in assessing the status of that species. At least to the federal judge hearing the case, the argument was sound; he ruled that NMFS regulators "'arbitrarily and capriciously' treated the... wild and hatchery salmon differently when listing the wild fish as threatened" (Verhovek 2002). NMFS has since been reevaluating the status of all of the listed ESUs of pacific salmon to include a consideration of hatchery fish.

Not surprisingly, environmentalists cried foul and appealed the judge's order; the result is neither surprising nor inconsistent with an approach to conserving biological diversity that is too heavily focused on species. Assuming the fisheries biologists are correct that the hatchery fish are not biologically, physically, or genetically distinct from the wild ones, then the wild and hatchery salmon really are the same species and it seems reasonable that both should be counted in assessing the status of the species. Does this mean that restrictions on activities detrimental to the health of the river and thus to the viability of wild salmon should be eased? Of course not. Rather, it means that we must expand our consciousness beyond species boundaries if our goal is to protect native biodiversity, the true intent of the Endangered Species Act.

What this case helps to show is that while keeping our focus on ESUs is necessary to an acceptable resolution of the salmon crisis in the Northwest, it is far from sufficient. In addition, we must resist an excessive species focus again to recognize and protect the larger, ecosystemic and evolutionary processes essential to any viable solution to salmon restoration and recovery. This more holistic perspective, looking beyond species, is necessary to understand both the causes of the current crisis and the answers that we must see. A first step, as Jim Lichatowich (1999) suggests, is to see that the issue is much bigger than the hydroelectric dams that choke the Columbia and other Northwest rivers.

> It is the cumulative effects of many activities in watersheds, not just the highly visible dams, that ultimately are responsible for driving the salmon

> toward extinction. This means that there is no simple solution—no single problem that, if attended to, will bring about recovery. The entire chain of habitats, from headwaters to the ocean needs attention. Restoration must address entire watersheds, or, to use the terminology popular today, we must begin managing entire ecosystems—managing human activities with ecosystem health in mind. (203)

We must, that is, look beyond any individual unit of biodiversity to a more holistic approach to our conservation.

The moral I draw from the Pacific salmon case is that we must think beyond species to achieve effective conservation of biodiversity. We must think smaller—to populations and evolutionary units—if we wish to safeguard the process of evolution and the genetic diversity upon which species, and ultimately all of nature depends. But we must also expand our consideration outward—to ecosystems and managing the human impacts, which threaten their health and vitality—if we wish to protect species. We cannot, as Lichatowich so powerfully argues, have *Salmon Without Rivers*. Nor, I believe, can we have any effective protection of biodiversity without this larger, more holistic approach to conservation, an approach clearly intended by the ESA.

How, I wonder, might these lessons apply to the case of wolves in New England? Can our predator conservation efforts be pushed beyond species—both upward and downward—in the way that I have argued is essential for conserving biological diversity? I believe they can, but as with salmon and other species, it will require a change in perspective, in the questions we are asking and the concerns on which we focus. It will require, in short, that we move beyond an excessive concern over the specialness of species.

The first step must be to deemphasize taxonomic concerns over what species of wolf populated this region in the past. Whether gray, red, *lycaon*, or some hybrid, the eastern wolves certainly formed an evolutionary unit distinct from populations in other parts of North America. Rather, our focus must shift from taxonomic to ecological questions. What role did wolves play in the Northern forest ecosystems? How were assemblages of biodiversity different when this pack-hunting carnivore occupied the top spot in the food chain? How have ecosystem health and integrity been affected by the changes wrought in the century of wolflessness in the Northeast?

Perhaps the largest and most important questions to be answered concern the role of the coyote as the new "top dog" in the forests of New England. Are the Eastern coyotes, who moved in to fill the ecological void left by the extirpated wolves, a sufficient replacement? The answer is unclear but perhaps holds the key to resolving the debate over reintroduction of wolves. We must also look beyond the questions of reintroduction to the barriers to natural recolonization by other populations of wolves, especially the population in Canada's Algonquin Provincial Park. The truest answers to all of our questions rest with the animals themselves and their actions, not ours. If we cannot have salmon without healthy rivers, neither can we have wolves without healthy forests. Here perhaps is the key difference between wolves and their coyote cousins, an ecological difference more significant than the taxonomic rules by which we draw species boundaries. While coyotes are generalists and opportunists, able to survive and even thrive in a humanized and damaged landscape, wolves are not, or at least appear not to be. Though we know that we can have coyotes without a healthy and diverse landscape, the same is almost certainly not true for wolves, of whatever species or kind. Should we reintroduce wolves to the Northeast? Perhaps, but an ecological perspective would, I think, shift our focus to protecting and healing our fragmented habitat before we seriously contemplate such an action. We need forests first, healthy ones worthy of their former hunter, and then perhaps wolves will return on their own, signaling to us that the time is right. But even this conclusion is too simple, since the wolves themselves can help to heal and improve their own habitat as we saw when wolves were reintroduced to Yellowstone National Park.[4] The point is that we must move beyond bound excessive focus on boundaries—taxonomic, geographic, political—in order to see what is right for wolves, humans, and conservation in general.

Beyond Boundaries

Expanding our conservation efforts beyond species to all levels of diversity is just one small, though significant, step in a broader shift in thinking. It's a shift to a perspective that recognizes the necessity of drawing lines and classifying, while preserving a view that extends beyond boundaries. The effects of this shift will extend well beyond the endangered biota of the "more than human world" to our vision and awareness of the human species and our thinking about the world around us.

The drawing of boundaries, like many forms of reductionism, is inevitable, and boundaries help us to understand, relate to, and protect the abundant diversity of the natural world. Though inevitable, boundaries are also dangerous and must be applied carefully. We must remember that these are *our* boundaries—yes, they are grounded in the subject of study and concern—but humans are the ones who draw the lines in the sand that divide up the immense natural diversity. We are not so much cutting nature at the seams, to use Plato's old metaphor, as drawing lines in the sand, lines that will wash away and be redrawn differently as our understanding and objectives change.

The drawing of boundaries is a human action for human purposes. In fact, the very ideas of classification, division, and hierarchy are human ideas. While these ideas, these concepts, are indispensable to human thought in general, and science in particular, they are also dangerous. They are dangerous because they can lead us to think that reality *is* our boundaries, we can conflate our aids to understanding reality with reality itself. The danger, in short, is that we mistake the map for the mountain. We forget that our tools for understanding are just that, tools not reality. While acknowledging the human element and purpose in the boundaries we draw is dangerous, it is also freeing. We can have good purposes as well as bad. We *must* have good purposes instead of bad.

Boundaries are an essential part of rational thought, but if we aren't careful we just might lose more understanding than we gain through the boundaries we draw, since there is more to the world than our rational minds can know or understand. There are, as Paul Rezendes (1998) and others remind us, "some aspects of life that thought cannot understand" since "thought works by compartmentalizing, creating boundaries—dividing the whole into parts" (14). Rezendes teaches his students of animal tracking that it is only by going "beyond thought" that we are able to fully comprehend the meaning that we find in the natural world (14). To understand deeply and fully many elements of our lives and the world around us, we must at times go beyond thought to the realms of mystery and perception where words and boundaries are equally alien.

So, too, with our relationship to biodiversity. We must classify and name, divide and rank; we must, in short, impose boundaries on the world in order to understand it, in order for our minds to grasp the complexity. But in grasping the complexity we also reduce it beyond its true character, its true and deepest meaning. While boundaries are necessary, so are unbounded places and ways of being and knowing. We must learn not to expect to understand everything, for this is impossible. Rather, we must come to accept our role and place as creatures within a boundless world of diversity.

Notes

1. Some writers on species concepts draw a sharp distinction between species *taxa* and the species *category* (Mayr 1982, 253-254). Species taxa are groups of organisms (e.g., blue jay is a species taxon). The species category is the class of all species taxa. Both the category and individual taxa are designated by the term "species," on this view, though they are quite different entities. While I recognize the distinction, I do not make use of it here. Nor do I consider it particularly helpful in sorting out the issues associated with defining species. How one defines the species category will inevitably affect how one delineates species taxa. Likewise, often it is through the

process of classifying organisms into species taxa that we come to a fuller understanding of the species category. In this chapter I stick to the simpler, though less precise, language of simply talking about species recognizing that it refers to these two different, but inextricably linked, entities.

2. While I have lumped all together here, there are many different phylogenetic species concepts with subtle but significant differences between them. For a good survey of the main phylogenetic species concepts with explanation of their distinguishing features see Wheeler and Meier 2000.

3. See Ereshefsy 1992 and Mishler and Donoghue 1982 for a review of the literature showing differences between competing taxonomies.

4. Thanks to an anonymous reviewer for pointing out the important link between wolves and habitat recovery in the West. The wolves in Yellowstone have altered populations of other species, particularly prey species, in ways that have changed vegetative patterns, contributing to a positive feedback loop that improves ecosystem health. At what stage of forest health a similar dynamic can be expected to occur in the Northeast is very difficult to predict, but we should definitely not wait until all ecological issues have been resolved or until we have perfect habitat before we reintroduce wolves.

References

Brewster, W. G., and S. H. Fritts. 1994. "Taxonomy and Genetics of the Gray Wolf in Western North America." In *Ecology and Conservation of Wolves in a Changing World*. Edited by L. N. Carbyn, S. H. Fritts, and D. R. Seip. Canadian Circumpolar Institute, Occasional Publication No. 35, 375–397.

Ehrlich, P. R., and P. H. Raven. 1969. "Differentiation of Populations." *Science* 165:1228–1232.

Ereshefsky, M., ed. 1992a. *The Units of Evolution: Essays on the Nature of Species*. Cambridge: MIT Press.

———. 1992b. "Eliminative Pluralism." *Philosophy of Science*, 59: 671–690.

Kitcher, P. 2001. *Science, Truth, and Democracy*. Oxford: Oxford University Press.

Lewontin, R. 2000. *The Triple Helix: Gene, Organism, and Environment*. Cambridge: Harvard University Press.

Lichatowich, J. 1999. *Salmon Without Rivers: A History of the Pacific Salmon Crisis*. Washington, DC: Island Press.

Mayr, E. 1942. *Systematics and the Origin of Species from the Viewpoint of Zoologist*. New York: Columbia University Press.

———. 1970. *Populations, Species, and Evolutions: An Abridgement of Animal Species and Evolution*. Cambridge: Harvard University Press.

———. 1982. *The Growth of Biological Thought: Diversity, Evolution, and Inheritance*. Cambridge: Belknap Press.

Miller, S. A., and J. P. Harley. 1992. *Zoology*. Dubuque: Wm. C. Brown.

Mishler, B. D. 1999. "Getting Rid of Species?" In *Species: New Interdisciplinary Essays*. Edited by Robert A. Wilson. Cambridge: MIT Press.

Mishler, B. D., and M. J. Donoghue. 1982. "Species Concepts: A Case for Pluralism." In *Conceptual Issues in Evolutionary Biology*. Edited by Elliott Sober. Cambridge: MIT Press, 217–232.

National Research Council. 1995. *Science and the Endangered Species Act*. Washington, D.C.

Noss, R. F. 1999. "A Citizen's Guide to Ecosystem Management." Biodiversity Legal Foundation Special Report, Boulder, Colorado.

Nowak, R. M. 1995. "Another Look at Wolf Taxonomy." In *Ecology and Conservation of Wolves in a Changing World*. Edited by L. N. Carbyn, S. H. Fritts, and D. R. Seip. Canadian Circumpolar Institute, Occasional Publication No. 35, 375–397.

Rezendes, P. 1998. *The Wild Within: Adventures in Nature and Animal Teachings*. New York: Berkeley Books.

Roy, M. S., E. Geffen, D. Smith, E. A. Ostrander, and R. K. Wayne. 1994. "Patterns of Differentiation and Hybridization in North American Wolflike Canids, Revealed by Analysis of Microsatellite Loci." *Molecular Biological Evolution* 11(4):553–570.

Ruse, M. 1987. "Biological Species: Natural Kinds, Individuals, or What?" *British Journal for the Philosophy of Science* 20: 97–119.

Starr, C., and R. Taggart. 1989. *Biology: The Unity and Diversity of Life*, 5th ed. Belmont, CA: Wadsworth.

Van Valen, L. 1976. "Ecological Species, Multispecies, and Oaks." *Taxon* 25: 233–239.

Verhovek, S. H. 2002. "'Saving' Wild Salmon's Bucket-Born Cousins." *New York Times* (February 4).

Wayne, R. K., N. Lehman, and T. K. Fuller. 1995. "Conservation Genetics of the Gray Wolf." In *Ecology and Conservation of Wolves in a Changing World*. Edited by L. N. Carbyn, S. H. Fritts, and D. R. Seip. Canadian Circumpolar Institute, Occasional Publication No. 35, 399–407.

Wheeler, Q. D., and R. Meier. 2000. *Species Concepts and Phylogenetic Theory: A Debate*. New York: Columbia University Press.

Wilson, P. J., S. Grewal, R. C. Chambers, and B. N. White. 1999. "Genetic Characterization and Taxonomic Description of New York Canids." In Paquet, P., J. R. Strittholt, and N. Staus. *Wolf Reintroduction Feasibility in the Adirondack Park*, A Report to the Adirondack Park Citizens Action Committee and Defenders of Wildlife. Corvallis, OR.

Chapter 5

Respect for Experience as a Way Into the Problem of Moral Boundaries

Charles S. Brown

Ever since Aldo Leopold[1] mused that extending the boundaries of the moral community to include the land was both an evolutionary possibility and an ecological necessity, environmental thinkers have searched for principled reasons to justify locating the boundaries of the moral community in new and different places. Efforts to redraw the boundaries of the moral community have frequently centered around the general thesis that core elements or beliefs within our worldviews have served to legitimate and encourage our reckless domination of the natural world resulting in the massive harm to nature known widely as the environmental crisis. According to their specific diagnoses, various thinkers have offered suggested cures involving some kind of revolution in thinking that would produce the kinds of attitudes and moral commitments needed to develop and sustain socially just and environmentally benign practices.[2] Common to this entire genre of philosophical analysis is the claim, or perhaps hope, that by identifying the elements in our worldview that are responsible for ecological destruction, it may be possible to develop alternative and ecologically benign worldviews that would free us of the seemingly historical inevitability of the mass production of destructive technologies that undo, rip, wound, and tear away at the biotic structure of the natural world.

A well-known version of this strategy, widely associated with the Deep Ecology movement, argues that it is the anthropocentric character of our traditional worldview that is largely responsible for ecological destruction and exploitation. According to this line of thought, our traditional anthropocentricism must be replaced with a biocentric or ecocentric worldview that extends the notion of intrinsic value, traditionally limited to humans, to all ecological forms and structures. Ecofeminists, on the other hand, have argued that environmental domination results not from an anthropocentric worldview but from an androcentric worldview that reduces nature to a "feminine other," thus capturing nature within its project of masculine domination.[3]

The project of unmasking ecodestructive elements in our worldview does not end, however, with the development of new, alternative, and ecofriendly worldviews, but rather in the more radical possibility of shifting power within our worldviews away from the controlling power of fixed concepts and categories and toward an openness to the manner in which the world unfolds. My defense of this claim is grounded in an interpretation of the promise and the legacy of the kind of phenomenological philosophy that has dominated a good bit of continental thinking throughout the twentieth century. One of the primary achievements of phenomenological philosophy has been the steady unmasking of the pretensions of metaphysical concepts and abstractions that serve as cornerstones or building blocks of any worldview.

For the most part, our thinking has been directed by sets of concepts and categories that are external to thinking itself. Our various concepts, understandings, and attitudes toward nature are deeply influenced, some might even say predetermined or prefigured, by historically constructed concepts and categories. To attempt to think without a radical questioning of the historical and contingent nature of the concepts and categories controlling thought is simply to articulate the combinatorial possibilities of fixed semantic regimes. Rather than give in to this prepackaged manner of thinking, we must hold out for a kind of thinking that is open to the world, a kind of thinking that is able to take the world in, to be available to the revelation that the world may offer. Such thinking accepts what the world offers. At its best, I am referring to what may be described as rational insight and, at worst, a kind of cheap mysticism. Such thinking would be characterized by its intrinsic revisability in the face of an always open future.

If we reflect on the basic impulse of Edmund Husserl's original phenomenological philosophy and the subsequent development of that tradition, we find a steady critique and unmasking of the taken-for-granted status of concepts and categories of the reductive metaphysical naturalism that results from Descartes's privileging of extension as the metaphysical essence of matter and nature. Husserl argued that such naturalistic metaphysics was essentially a mass appropriation of culturally constructed, idealized, and abstract objects of a mathematized physics, which purports to be not only a faithful representation of reality itself, but the only possible one. By the construction of scientifically respectable, measurable properties as "the real" and rationality as "scientific method," reason has become trapped in the success of its own natural sciences. With the further interpretation of rationality as "value-free," reason losses the ability to confront problems of value. Husserl's reaction to reductive metaphysical naturalism helps us to see the consequences of a view of nature consisting entirely of extensional properties externally related to each other within a causal matrix. Reason becomes computational and instrumental at best, and nihilistic at worst. Such a value-free mechanistic conception of nature inevitably leads to moral, social, political, and ecological crises as the value-free conceptions of rationality supporting such a naturalism dismiss the good as mere subjective preference, thus removing all questions of value from rational discourse. Such a dismissal of the good from the real and the rational generates intractable problems for moral philosophy in general and environmental philosophy in particular.

Phenomenology's specific contribution to ecological philosophy begins in an attitude of respect for experience that it shares with much of ecological philosophy and many environmental activists in general. Unlike naturalism, phenomenology does not seek to dismiss experience as subjective, nor does it wish to replace or reduce experience to a more fundamental or more basic mode of being. Phenomenological description and articulation of the structures of experience are an attempt to, as Husserl puts it, return to the "things-themselves,"[4] rather than simply taking for granted higher-level, culturally sedimented idealizations and abstractions that often pass for ahistorical metaphysical discoveries. Such attention to and respect for the way the structure and meaning of our involvement with the world unfolds within everyday experience, and thus is the great stuff of phenomenological description, provides a kind of corrective to the kind of thinking controlled by a worldview—that is, the kind of thinking that always reinterprets ordinary experience according to the concepts and categories occupying positions of power within such a worldview.

Worldviews dominated by anthropocentric assumptions have long informed a great deal of thinking in modern and premodern culture. Indeed most Western thought has been blatantly anthropocentric as both Enlightenment political philosophy and monotheistic religion are deeply anthropocentric. Supported by Cartesian dualism, this widespread, deep, and largely unquestioned element in our worldview often controls and guides what passes for rational thought. Its natural tendency is to mold, shape, and reinterpret the content and structure of everyday thought according to its own self-image. Being under the seductive and self-serving sway of anthropocentrism we regularly distort or even dismiss basic and fundamental structures of the way the world unfolds before us, thus misinterpreting the meaning of our experience.

Recall the stories about the so-called natural scientists who, under the sway of the anthropocentrism of Cartesian dualism, performed vivisection on animals, while interpreting their screams and howls not as genuine expressions of suffering and distress but as mere unmeaning mechanical responses. Surely, if we look at and listen to such screams and howls, not through the distorting lenses of anthropocentric metaphysics, we are immediately confronted not with mechanically produced sound and motion, but with the immediate and natural expression of pain and suffering. We may also be confronted with the immediacy of a moral imperative or, at least, the irreducible sense of a vague and diffuse moral unease.

This basic tendency to dismiss or even falsify the character of ordinary experience can also be seen in experience closer to home. Many of us are deeply and empathetically intertwined with the lives of our companion animals. As lap dogs, large or small, lie on our laps and housecats rub their bodies against our ankles, we feed them when they are hungry and comfort them when distressed; in short, we recognize the dignity and integrity of their own good as we take up their good as our own. The basic tenor and feel of such experience, the basic categorial structure of such experience, are decidedly not anthropocentric. Such experience does not register the boundaries of moral concern with the human community. Such experience does not legislate a priori boundaries of genuine or illusory moral experience. It remains open to the way in which the world unfolds.

Our companion animals are regularly experienced as having something like what the philosophers call intrinsic value. They are certainly experienced as having their own good. But the import of the meaning of such experience is regularly denied or dismissed in moments of bad faith as we systematically overlook the import of such experience and interpret it as mere subjective sentiment. The recognition of such phenomena points, I hope, to a kind of thinking in which the balance of power between our historically sedimented a priori constructs of the world and our authentic and immediate openness of the world are realigned in favor of the latter.

A bomb-sniffing dog was killed in the attack on the World Trade Center towers. His remains were subsequently recovered in the excavation of that site, and later he was given a massive tribute by the K-9 community in New York City. The national media quoted his trainer or handler as saying, at the dog's funeral (an elaborate and formal ritual), something very much like the following: "To many people he was a dog, but to me he was a friend." We all understand such an expression even as we dismiss it as simply "sentiment." We dismiss and ultimately disrespect such experience when we do not claim it as our own. We let such experiences float freely as mere epiphenomena. We enjoy the content of such experience, but let it go before the import and implication of such experience make their way to the center of our lives and our experience. As we typically do not claim such experiences as our own we enjoy them momentarily and then dismiss them as promiscuous lovers would. Such dismissed experiences never gain the kind of authenticity[5] needed to serve as a ground of a more generalized kind of non-anthropocentric thinking.

Just as anthropocentrism encourages us to dismiss the meaning and import of much of our pre-reflective experiences involving nonhuman animals, reductive naturalism and the eliminative materialism that it implies go even further and encourage us to misrepresent or falsify the meaningful and value-laden manner in which we experience ourselves and other people. The further claim to be argued here is that reductive naturalism not only dehumanizes humans and mechanizes animals, but also that thinking under the sway of reductive naturalism dismisses natural properties of goodness we experience as inherent in the natural system of the Earth's biotic web as well.

Within our pre-reflective experiences, we regularly find the world and the things within it to be infused with value. The sun, the rain, and all manner of others are regularly experienced as good. Our everyday life is filled with moral sentiments that appear from a phenomenological perspective as instances of a pre-reflective axiological consciousness, that is, as an intentional and evaluative aiming at objects and states of affairs. Our various understandings of the good are, however, open to continual reassessment in light of subsequent experience, just as we continually reassess our previous understandings of the real or the true. In this way, value experiences exhibit their own kind of objectivity and, as in the case with perception, any one experience is given as provisional and revisable in light of future experience. In this fashion, moral consciousness exhibits its own kind of objectivity and its own kind of rationality, grounded within the dialectics of empty and filled intentions.

To say that we directly experience the rain, sun, and soil as good is not simply to say that we enjoy them. The experience of finding something to be good and finding something to be pleasurable are not the same. Pleasure is an immediate quality that does not refer beyond itself. The good is experienced, in part at least, as a referential structure that implicates or anticipates its own fulfillment in future experiences. When we experience friendship as good, we have a sense, even if unarticulated, of why or how what we experience as good is good. Even if we cannot express it, we know that friendship extends our sphere of concern while comforting us in ways that provide our lives with meaning. Value experiences, with their inherent sense of anticipation, bring with them their own procedure for confirmation. Within the intentional structure of experience, including axiological experience, lies a recipe or an inner logic that provides or denies justification of the lived sense of that experience. Our experience of water, soil, or friendship as good is always the experience of a prima facie good, never as a final and absolute good. For example, to experience friendship as good is to interpret and impose the sense of good on friendship, but it is also to expect to continue to find goodness in friendship and to have such expectations fulfilled. The very experience of friendship as good is bound up with an implicit understanding of the implications and, in short, meaning of friendship. Our experience of the goodness of water, soil, air, and friendship is always subject to the possibility of breakdown.

The empiricists have similarly attempted to ground morality in experience, as I am suggesting here, and Hume's efforts along these lines were certainly the most perceptive, rightly pointing to the importance of moral sentiments. Without a theory of intentionality, however, and operating with an impoverished concepts of intuition (the theory of ideas), he could not get beyond the subjectivism that later ended in emotivism, an emotivism, generally thought to be grounded in that very form of naturalism that separates the real from the good. By the very separation of the real and the good, ordinary experience is disrespected, and our experience of the good is judged to be subjective, personal, private, and reducible to a kind of nonrational preference for pleasure at best and to the stimulation of brain processes at worst.

Value experiences have their own value horizon. To experience something as good is to know what to expect of it and to have an implied sense of its anticipatory structure. We experience rain, sun, and soil as good not simply because we enjoy or find pleasure in them but because we appreciate their roles in sustaining the Earth's biotic web. The better we understand the role water plays in the biotic fabric of living nature, the more sophisticated our appreciation of water becomes. When I quench my own thirst with a cool drink of fresh and clean water, my sophisticated or simple appreciations of water are fulfilled in a moment of bodily awareness of the goodness of water. We experience sun, rain, and soil as good when we glimpse their roles in the fabric of the planet's biotic web. As we experience our own dependency on the planet's biotic web, we realize the massive and inescapable interdependency among other species and processes in a mutually sustaining web of life. As we are constantly faced with threats to clean air and water and with the possible loss of those goods, we become more open to the possibility of experiencing air, water, and soil in a rational axiological categorial manner.

It is within this context that the moral horizon emerges, within the context of the human situation, within the context of our experience. We are biological evolvents, existing within the biotic web as well as self-conscious subjects of that life moving forward toward our deaths while embracing and valuing the life we live. Our pre-theoretical experience, infused with cognitive, evaluative, and volitional moments, is not the experience of an "objective world" (i.e., of a devalued world consisting of causal relations and extensional properties), but rather the directly lived, laden with value and meaning. It is this meaningful order, constituted by the presence, activities, and function of life, that provides the deep context for the emergence of moral experience.[6]

This meaningful order does not have the status of fact. It is not a "given" of experience, but rather, to use a Husserlian locution, it is "pre-given," or to use another phrase popularized by subsequent phenomenologists, it is "always already" there. This meaningful order of life, this ecology of bios, within which we are experientially intertwined, is the experiential ground of our intuitions about holism, as well as a condition of the possibility of moral consciousness. This meaningful order of purpose and value is part of the unnoticed background of experience available for philosophical reflection. There is every reason to believe that this meaningful background of purpose and value has existed long before the human species, and that our specifically human goods only exist within a larger system of good arising from the biotic and prehuman awareness of nature as good. This awareness, arising from the experience of being a part of and dependent on that order, is perhaps the source of age-old intuitions that goodness itself is beyond humanity.

It is this meaningfully ordered and value-laden world of our direct experience, available to us in pre-metaphysical experience of the world, that ultimately justifies all moral claims.[7] We know that dishonesty, fraud, rape, and murder are evil because they each, although in different ways, retard and inhibit the intrinsic purposes and desires of life, which presents itself as a value for itself in our most basic and world-constituting intuitions. Value experiences occur within a meaningfully ordered value horizon. It is this value horizon of life and our ongoing validation of the good that supplies the final justification of our experiences of the good. It is within this value horizon of life that our experiences of good and evil are shown to be more than "mere subjective preferences." If we initially find friendship to be an evil and fraud to be a good, an openness to further experience will almost always correct this. Finding value in friendship and disvalue in fraud is not arbitrary.

In our everyday experience of value, we regularly find food, clothing, shelter, community, and friendship as good. Rarely do these things disappoint us. Our experience continues to establish these as goods in an ever-evolving process of being open to the good. By grounding ecological philosophy in the evolving wisdom of our collective experience, we can avoid the twin perils of absolutism and relativism. We avoid dogmatic absolutism by understanding that our experience and conception of the good are always open to revision, and we avoid relativism by recognizing that our experiences of the good demand their own confirmation in future experience. It seems to be a fundamental possibility that humans can and often do experience nature as infused with goodness and from within an attitude of concern and empathy. That moral unease that underlies much of the public awareness of environmental crisis is a symptom of that possibility. That moral unease may not be the full-blown deep ecological intuition of which Arne Naess and others speak, but it is an authentic moral sentiment leading us to an attitude of concern for nonhuman value that can only be described as moral.

While our current configuration of technocentrism and consumerism, structured by a nihilistic rationality with anthropocentric assumptions, may encourage the dismissing and disrespecting of such experiences, growing numbers of people continue to experience the ecological crisis as a moral harm done to the goodness of nature and the Earth itself. This is simply to say that experiencing the events of planetary destruction and waste that constitute the ecological crisis is increasingly recognized and authentically claimed as a morally charged experience for many people. Of course, such non-anthropocentric experiences could not and cannot be expressed within traditional anthropocentric moral discourse and with the instrumental value-free rationality that usually supports it. We can read a great deal of nature writing, from Thoreau and Muir through Leopold and on to today's Radical Ecologists, as attempting to establish a new mode of moral and aesthetic discourse in which experiences of the intrinsic goodness of nature can be registered, expressed, and rationally developed. Without such a vehicle of articulation, experiences remain mute and powerless and are dismissed to the margins of rationality.

Now, a certain kind of axiological unease pervades a growing number of people's experience of ecological destruction and change. The environmental and ecological changes brought about by industry, mining, and overconsumption are no longer simply seen as necessary by-products of the conversion of raw material into consumables, but such changes are now regularly experienced as a moral harm to the nonhuman natural world. Sadly, such experiences are informed by a haunting vision of the Earth's wounds and irrevocable tears in the biotic web as well as growing systems failures. One of the essential characteristics of any web of life is that the web itself and the co-evolvents within the web are subject to the possibility of death. The very idea of the Earth's mortality helps to explain the urgency in the call for an ethical response that the experience, direct or otherwise, of the growing ecological disaster solicits. Our task now is not only to recognize and claim such experiences as our own, but also to cultivate a kind of thinking that respects the basic and fundamental way the world unfolds before us. Such thinking is a kind of bottom-up thinking in which we find and respect a measure of rationality internal to experience itself.

Notes

1. Aldo Leopold, "The Land Ethic." In *The Sand County Almanac*. New York: Oxford University Press, 1949.

2. See Michael Zimmerman, *Contesting the Earth's Future: Radical Ecology and Postmodernity* (Berkeley: University of California Press, 1994) for a critical review of the varieties of radical ecological thought.

3. Charles S. Brown, "The Real and the Good: Phenomenology and the Possibility of an Axiological Rationality." In *Eco-Phenomenology: Back to the Earth Itself*, ed. Charles S. Brown and Ted Toadvine. Albany: State University of New York Press, 2003, 3.

4. Edmund Husserl, *Logical Investigations*, vol. 2, trans. J. N. Findlay. New York: Humanities Press, 1970, 252.

5. My thinking here seems obviously influenced by both Heidegger's analysis of authenticity in *Being and Time* as well as Sartre's description of bad faith in *Being and Nothingness*.

6. Similar points are made by Erazim Kohák in a variety of publications. See especially his essays, "Knowing Good and Evil" in *Husserl Studies* 1 (Summer 1993): 31–42, "Varieties of Ecological Experience" in *Environmental Ethics* 19 (Summer 1997): 153–171, and "An Understanding Heart: Reason, Value, and Transcendental Phenomenology," in *Eco-Phenomenology: Back to the Earth Itself*, ed. Brown and Toadvine.

7. Charles S. Brown, 2004. "The Intrinsic Rationality of Moral Phenomena," in *Skepsis* XV/i: 477–494.

Chapter 6

Boundarylessness: Introducing a Systems Heuristic for Conceptualizing Complexity

Beth Dempster

Imagine a deep, thick West Coast temperate rainforest. Massive tree-trunks reach upward; branches interconnect in a three-dimensional weave that blurs the distinctions among individuals at the canopy level. Lush green undergrowth covers the forest floor. Filtered sunlight, caught by broad leaves, pumps surfeit water. Rotten logs provide a vertical head start to species that can accommodate the acidity. They also provide hiding places for a multitude of small mammals, amphibians, bugs, worms, and other creatures. The deep silence belies a frenzy of activity. Sometimes—if I'm quiet—I imagine I can hear trees growing and soil microorganisms moving about. A fascinating, vibrant, complex ecosystem.

Yet the complexity only *begins* with such a biophysical description. Because I grew up in British Columbia—with a resource-based economy reliant on forestry—the economic and social importance of forests has been readily recognizable to me. Integrated into the forest ecosystem are loggers, silviculturalists, researchers, and buyers; the clean fresh smell of just-cut lumber; the fine grain of old-growth smoothed under a carpenter's hand; the steaming pile of meat and potatoes served as Sunday dinner in a small town; the barren stumps and slag piles of a clear-cut perceived as a blight, a new beginning or a means for providing that Sunday dinner, depending on the position and background—and species—of the viewer. The complexity, interconnections, and interdependencies among the ecological, social, economic, ethical, and other aspects seem to defy understanding.

Conceptualizing this situation in a manner conducive to manifesting sustainability—of ecosystems, economies, cultures, and people—continues to be a significant and increasingly important challenge. While completing an undergraduate degree in forestry and wrestling with a small corner of this challenge—understanding the biophysical sustainability of forest ecosystems—I found (and still find) appropriate concepts and heuristics to be lacking. Ecologically, I learned about large-scale "experiments" estimating nutrient and hydrological inputs and outputs over entire natural and harvested watersheds. Biologically, I learned about the requirements of different species for soil moisture, soil nutrients, and shade tolerance. Economically, I learned about nonmarket contingent valuation techniques as a means to "value" the invaluable. Ethically, I debated the challenges and implications of stretching the boundaries of moral communities beyond humans. This diversity of concepts and ideas provided valuable understanding, yet they relied on categorization, reduction, and quantification. Complexity was simplified by isolating parts and drawing boundaries; by making distinctions: forest/nonforest, in/out, good/bad. While useful, these notions and requirements fit uncomfortably with my experience.

In particular, my search for concepts that would enable the characterization of ecosystems without requiring boundary delineation was unsuccessful. My response—through the naive intentions and desperate creativity of a fourth-year student finishing her undergraduate thesis—was to invent a word and the subsequent description to occupy the void: *sympoiesis* (Dempster 1995). The term was developed from the Greek for *collectively producing* and was chosen as a contrast to *autopoiesis* (Maturana 1980a, 1980b, Maturana and Varela 1987), Varela et al. 1974,). In brief, sympoietic systems are characterized by complex interactions among components and relations, which recursively produce and maintain a self-similar, dynamic organization of evolving, interdependent complexity. This complexity is distinguishable as some sort of "whole" or entity, yet has neither temporal nor spatial boundaries. Rather than maintaining their "wholeness" through the delineation of boundaries, such systems are maintained by dynamic interdependencies and an interactive, balancing tension among components, processes, and influences.

For many, the most contentious aspect of sympoiesis is the notion of boundarylessness. People point to the obvious *existence* of boundaries—including skin, bark, borders, and cultural norms—to indicate the problem. I concede to the importance of questions about what a "boundary" is, and about how it might be characterized. However—regardless of how firm, fuzzy, permeable, variable, arbitrary, conceptual, useful or otherwise boundaries are taken to be—I resist statements that unreflexively claim their undeniable "existence." I speculate that such emphasis arises from our overriding dependence on the visual and physical and also on using categorization, "bounding," and reduction as means for understanding and interacting with that-which-exists. Agreeably, boundaries enable particular—and valuable—ways of understanding, yet I maintain an interest in considering how to conceptualize "systems" *without* relying on the delineation of boundaries. I see this as an *alternative to*, rather than replacement for, more conventional approaches—an alternative that may *also* lead to valuable ways of understanding.

In this chapter, I briefly describe the notion of sympoiesis as a heuristic to aid conceptualization of boundarylessness. I find such an approach critical in my work, which attempts to grapple with the multiple challenges involved in planning sustainability. Reflecting the demands of such work, discussion is framed and written for an interdisciplinary audience. As previously noted, my own background lies in forestry, ecology, sustainability and planning—with mere dabblings into the realm of philosophy. The inclusion of my chapter in this collection of philosophical work represents—to my mind—a serendipitous dissolution of disciplinary boundaries.

I begin with a few comments to place myself on an ontological-epistemological map. I do this not to align myself with any particular philosophical tradition, but simply to state my position. These comments are followed by brief discussions of systems and boundaries, both of which I take to be heuristics. Since the concept sympoiesis arose from a desire to relinquish boundaries, I outline some arguments against boundaries, although taking care to indicate that they also have value. I then describe the notions of autopoiesis and sympoiesis, with brief examples for illustration. The ideas and theories presented build on notions of complex self-organizing, self-producing systems. The essential reason for building the theoretical platform is that conceptualizing phenomena according to these heuristic offers opportunities for understanding situations without the need to delineate boundaries. The conceptualization of sympoiesis also motivates a search for richer metaphors to describe the interconnected and interdependent nature of social-ecological systems. The brief examples of system descriptions and brief reference to Deleuze and Guattari's metaphor, rhizome, are intended to be little more than tantalizing possibilities.

Reality

> There is no single reality, but rather multiple realities, and what is represented depends on one's position in the field of negotiation.... It is about an ongoing process of negotiating reality.
> —Elizabeth Bird

I consider myself an ontological agnostic—generally unconvinced about the certain existence or nonexistence of reality, and/or about the degree to which

I/we construct it. In manifesting, experiencing, and understanding daily life, a continuum of realities seems most reasonable. At one end is *physical reality*: I sit in chairs, bump into tables, and occasionally worry, when backpacking, about people falling off cliffs. At the other end is *constructed or abstracted reality*: I wonder about meaning, bump into social norms, and occasionally worry, when writing, about falling into reification. Between and around these ends are *negotiated realities*—trade-offs between the problematic extremities. Bird, quoted above, refers to an "ongoing process of negotiating reality" (Bird 1987, 258). While my reality might be constructed, there seem to be limitations to what I can construct, hence the need for negotiation.

While I make these as ontological claims, I acknowledge their political and epistemological implications. By drawing such distinctions, I imply that constructed realities such as social norms are more malleable and less solid or harmful to bump into[1]—yet the solid walls of social norms can be as hard and immovable as solid rock. In a similar vein, it is crucial to recognize that physical reality is only "real" because I am a physical being. Neutrinos and ghosts do not bump into tables.

Fundamental to the process of negotiating reality is the need for, and use and development of, instruments and heuristics for conceptualizing those realities. We rely on tools and concepts to gather and organize our perceptions and sensations into understandings and ideas that enable our capacity to learn and to move successfully through our lives. Yet these tools and concepts themselves arise from our negotiated realities. This raises the question: How much do our heuristics, our interpretations of reality, and our realities influence the "negotiation" each of us engages in? How much do the tools and schema we use for *conceptualizing* reality *determine* our "reality"? These "how much" questions are unanswerable, yet they point to the necessity of attending to the weight and degree of—or at least acknowledging the presence of—the influences involved in developing our understanding. For example, to be specific with respect to boundaries: How much does our reality have boundaries simply because we use boundaries to conceptualize our reality?

Systems

One of the heuristics I find valuable for conceptualization—both theoretically and aesthetically—is the concept of "system." While some will argue that systems (just like boundaries) *exist*, I follow other systems-thinkers and consider them as heuristics: "As any poet knows, *a system is a way of looking at the world*. The system is a point of view—natural for a poet, yet terrifying for a scientist!" (Weinberg 1975, 52). Decreasing emphasis on Weinberg's use of sight, I consider a system to be a way of *conceptualizing* the world. Systems are heuristics that enable organization of perceptions and sensations by drawing relationships among them to conceive them as *wholes*—as entities or sets of things that somehow belong together. A system is taken to consist of parts interacting to form a whole. This whole is embedded in an environment, with which it exchanges inputs/outputs. Systems can have greater and lesser degrees of complexity—such as the difference between organisms (organic systems) and mechanical systems. Complexity may arise from a variety of factors, including hierarchy, feedback, self-organizing influences, or from the presence of different *types* of parts. While such complexity calls out for reduction to gain understanding, the key characteristic of a systems approach is *not* to do this, but rather to take a holistic approach.

In my early research, while trying to understand the "sustainability" of West Coast rainforests, I found these characteristics helpful but also inadequate. In particular, boundaries are typically taken as a defining characteristic of a system, yet I balked at the instruction: "Draw a boundary." Regardless of how temporary, permeable, or fuzzy the boundary was drawn or how carefully, conscientiously or reflexively the drawing was done, such an approach did not accommodate the experience of forests I described earlier or my sense of the understanding required for their appropriate management. I subsequently turned to consider the possibility of relinquishing boundaries.

Boundarylessness

> Distinction is the essence of boundaries. To illustrate, consider that a system may be defined without explicit attention to boundaries. A system may be conceptualized in terms of a set of interrelations among components and actors. However as soon as a system is conceived, so is its boundary. Simply by including one set of actors and components in one's understanding of "the system," others are excluded. There is 'system' and there is 'not-part-of-this system.' There is a boundary. The boundary is the conceptual space where the distinction between system and not-system happens.
> —Martin Bunch.

Boundaries are generally taken as necessary for making the distinction between system and environment. While definition of boundary as a "conceptual space" could provide another approach for alleviating some of my boundary-related concerns (noted later), I question the clarity—and the necessity—of the distinction articulated in this quotation. System versus not-system indicates a binary conceptualization of the world, which—although popular—is not the only one available. Alternatives to such binary logic are numerous. As an example, Janian logic has seven instead of two categories and three of them—is, is not, is indeterminate—parallel an informal categorization I remember chanting in childhood: yes, no, maybe so. Think of the primary colors red, blue, and yellow or the gradation of a color wheel, rather than black and white. Consider the difference between "we" and "us versus them." While the latter binary categorization sets up a conceptual (and also often cultural, racial, gendered, etc.) boundary between two groups, "we" simply identifies a collective of individuals—a "system," if you will, but one that does not need to be defined by distinction against "other":

> "We" does not require an opposing "they," "we" also denotes the relationships between "you" and "me." Once the term "we" is understood communicatively, differences can be respected as necessary to solidarity. Dissent, questioning, and disagreement no longer have to be seen as tearing us apart but instead can be viewed as characteristics of the bonds holding us together." (Dean 1996, 8)

One of my concerns regarding the use of boundaries is their easy application toward the construction of binaries, which can be problematic by creating, for example, a false sense of certainty: With only two categories, one can be certain that anything *not* in one *is* in the other: if not white, then black; if not right, then wrong. Even the addition of one more category increases the complexity and uncertainty: if not blue, then maybe red or maybe yellow; if not yes, then maybe no, but maybe maybe. Gradations tend to raise questions and perhaps contingencies: Does that seem blue to you? Or kind-of greenish? Well, probably not right now, but maybe later?

A related concern is that the distinctions made with boundaries, especially binary distinctions, so often become categorically oppositional—with "boundary" enabling the opposition and leading to a territorial attitude. The focus shifts to: in/out, black/white, right/wrong; boundary maintenance becomes critical, boundary defense even more critical; differences gain central importance, and relations become deemphasized. In so many cases this seems to be misleading and potentially damaging.

> notice that the opposites of inside vs. outside didn't exist in themselves until we drew the boundary.... It is the boundary line itself, in other words, which creates a pair of opposites. In short, to draw boundaries is to manufacture opposites...every boundary line is also a potential battle line, so that just to draw a boundary is to prepare oneself for conflict. Specifically, the conflict of the war of opposites, the agonizing fight of life against death, pleasure against pain, good against evil.... The simple fact is that we live in a world of conflict and opposites because we live in a world of boundaries. Since every boundary line is also a battle line, here is the human predicament: the firmer one's boundaries, the more entrenched are one's battles. (Wilber 1985, 18–19)

This statement echoes some of my own concerns. Yet I also discern a glimmer of the challenge and paradox involved in relinquishing (or at least attempting to relinquish) boundaries—for the statement itself carries a hint of boundary/battle cry: *every* boundary line *is* a battle line.[2] Letting go of boundaries presents a continual and paradoxical challenge since—at least to some degree—any firm statement against boundaries sets a boundary. Relinquishing boundaries does not mean arguing against them (only), but *also* offering other means for conceptualizing that-which-exists and maintaining some flexibility regarding when/how either one is applied. No-boundary-thinking seems to be a fundamentally different *type* of thinking from boundary-thinking. Advocating the former, however, does not necessarily imply negation of the latter. For example, I am not adamantly *against* boundaries and boundary-thinking, I just believe that no-boundary-thinking—as the current underdog—needs greater encouragement.

The predominance of boundary-thinking raises questions about how entrenched the concept of "boundary" is to our way of thinking. I sense subtle influences here. It seems that boundaries are so fundamental to particular styles of thinking and ways of being that they are assumed; they are an ingrained influence on both our perception and our conception of that-which-exists. How many of us are programmed (predisposed?) to think: boundaries? How many of us are decidedly uncomfortable with (rather than just perplexed by or new to) the idea of boundarylessness?

As a further concern, I sense that relying on a boundary to delineate the separation of system from environment promotes a tendency to disregard the environment as background. The complexity of interactions, interrelations, and interdependencies that make up "the environment" are reduced to inputs and outputs or written off as insignificant and irrelevant rather than recognized as constitutive—and full of both uncertainty and possibility. Drawing a boundary enables not only reduction, but selective attention. The "system" becomes the central focus, with all else comfortably designated as quantifiable flows in and out and all but excluded from consideration. By designating it as "the environment," it often seems to me that we give ourselves permission to discard everything beyond the boundary from concern and attention.

Admittedly, I do not *know* that boundaries—even fuzzy, permeable, temporary intermittent boundaries—have this effect. But I wonder. I wonder because examples of such restricted focus and release of attention seem commonplace: stand-density diagrams for managing forests; the enclave mentality in protected area management; territorial displays over cultures, race, or academic disciplines; the conventional "scientific" approach to medicine; "success" measured by income bracket. Each of these is facilitated by the presence of a "boundary" to identify the subject of focus. Yet each of these examples points to concerns and behaviors many (now) consider problematic. Attempts to learn or create our way out of these situations can readily be found: greater park ecosystem conservation planning to incorporate factors outside park boundaries; hybrid cultures, antiracist rallies and interdisciplinary studies as more common activities; "holistic" medicine to heal body, mind, and spirit together; ads for cars or bank-loans that—while still suggesting the need for money—picture successful people relaxing by a lake, playing with their kids, or driving through a thick forest.

Such a shift can be conceptualized in at least two ways: as indicative of expanding boundaries or of relinquishing boundaries. Since the former externalizes or refocuses rather than addresses the concerns, I apply my efforts to develop and understand the latter. Relinquishing boundaries, however, is not to denounce valid reasons, preferences, or values in using them—sometimes in someplaces. The point is to consider the conceptualization of boundaryless systems as an *alternative*. As noted with respect to the quote from Wilber, to do anything else sets a boundary rather than relinquishing one.

As with systems, I take boundaries to be conceptual devices—heuristics for simplifying the conceptualization of a set of phenomena by encircling the "whole" that is sensed to exist. They are valuable for making clear and precise distinctions and clear and precise system identifications. Given the noted concerns, however, I believe there is also value in having other approaches for identifying and describing systems. Possibilities include the establishment of distinctions by characterizing and comparing system attributes or the delineation of systems by drawing relations rather than drawing boundaries: you and me and whoever else comes along to join us, rather than you and me against the world. "Note that a system is a whole thing, and although complex it has parts that are connected to each other in some way; thus smaller parts of systems can be identified, but it is the connection which makes it a system" (Chadwick 1966, 184).

It seems to me that the notion of "system" is extraordinarily well suited for addressing the type of situations in which boundary delineation is particularly problematic. For example, many entities or sets of phenomena—such as ecosystems, economic systems, and cultural systems—are quite comfortably and consistently referred to as "systems" despite their lack of obvious boundaries. Reference to social, knowledge, transit, information, or management *systems* is generally used to indicate their interactive, composite, relational nature as identifiable—and singular—entities; parts-in-relation that form a whole. In these cases, components are readily identified as "belonging together" without enclosing them within boundaries. Rather than being an essential characteristic defining the very nature of "system," it seems that boundaries are conveniences drawn to simplify the complex interconnected interactions that are readily recognized as "systems." Boundaries are attempts to accommodate our cultural and methodological preferences for contained, quantifiable, categorizable entities.

In the big, wild British Columbia rainforest that I am familiar with, for example, the *components*—such as trees, plants, and other living organisms—might reasonably be defined by drawing boundaries to make system-environment distinctions (although mychorizzal fungi provide a bit of a challenge). However, defining the *forest* in this manner seems inadequate. We draw boundaries to facilitate particular styles of scientific inquiry and management, yet standing among the tall trunks and green undergrowth, the overriding sense is not boundary, but relation and interaction. For me, it is the latter two qualities that characterize the very essence of "system." Insistence on boundaries seems to dampen the valuable potential for conceptualizing reality that "systems" offer. There seems benefit in disconnecting *per*ception from *con*ception. To facilitate the conceptualization of forests—and other such complex entities—as "systems," I propose relinquishing the need for boundaries as a defining characteristic. Instead, I propose placing emphasis on components and especially on relations as a means of interconnecting parts to conceptualize "wholes."

In the case of complex systems, I propose the notion of sympoietic systems: complex, boundaryless, collectively producing systems (Dempster 1995, 1998a), choosing the term in contrast to the concept of autopoietic systems—self-producing, composite systems that generate their own boundaries (Maturana and Varela 1980). Further detail on the notions can be found in earlier work from which some of the following explanations have been summarized (Dempster 1998a, 2000, 2003).

Autopoietic and Sympoietic Systems

In addition to more general conceptions of systems, there are a number of more specific concepts. As noted, I developed the notion of sympoiesis in contrast/addition to the notion of autopoiesis. To explain the heuristic autopoiesis, I describe organisms as autopoietic systems and then explain some of the theory. I note two caveats. First, this description relies on earlier work of Maturana and Varela (e.g., Maturana and Varela 1980), which has a biophysical basis—even when applied to social systems.[3] Second, as will be recognized in the following discussion, I am continually challenged by the process of negotiating reality and my own tendency toward reification. Whether the slippage evident in my writing reflects the way things *are*, the way I have *learned* them, or the way I *construct* them, is a question that remains—for me—continually outstanding.

We generally conceive organisms to be composed of interacting parts—such as cells and organs—that work together through a variety of processes to create a whole entity: a living individual. Any particular type of organism—tree, salamander, person, or mychorizzal fungi—is identified by a particular arrangement; a specific set of relations connecting specific parts and processes. A tree, for example, is identified by the way in which roots are connected to leaves through trunk and branches and the processes of transpiration and photosynthesis. This arrangement is referred to as their *pattern of organization*.[4] Somewhat different patterns of organization are recognizable in deciduous and evergreen trees, which have different types of leaf arrangements and vary in their capacity for transpiration and photosynthesis across seasons.

An organism survives and maintains itself through the exchange of material and energy with its environment. What counts as "food" is defined by the organism and is monitored and restricted by a boundary. To the degree possible, an organism will prevent inputs that would alter its components, its processes and/or its pattern of organization in a way that would disallow continued survival. The boundary—bark and leaf surface, for example—is produced and maintained by the organism to establish it as separate from its environment and to regulate exchange. The physical manifestation of the system—the actual entity that exists in the physical domain—is referred to as the *structure* of the system.[5]

A somewhat simplistic analogy that is useful for explaining the autopoietic-theoretical aspects of this description is to think about pattern of organization as an interactive dynamic blueprint, and structure as its interactive dynamic manifestation in some domain, such as the physical domain. Distinguishing between these two concepts enables conceptualization of two other notions.

Organizational closure refers to the degree of self-containment a system has with respect to its pattern of organization. A system can be organizationally *open*, *closed* or *ajar*,[6] depending on whether its pattern of organization is completely determined by *external* sources (such as a car), completely determined by *internal* sources (such as an organism as described earlier), or primarily determined by internal sources but also influenced by external sources (such as an ecosystem as will be described next).

Structural coupling refers to the interdependencies that exist (or emerge) between a system and its environment in the domain in which they both exist. For example, food is a structural requirement for an organism (manifest in the physical domain) that is obtained from its (physical) environment. A system's need to obtain specific food from a specific environment reflects the structural coupling between them.

These distinctions allow conceptualization of systems that are simultaneously open and closed, in different, mutually interactive, ways. Systems can be *organizationally closed* but *structurally open*. For example, a living organism with an internal genetic "program" determining its pattern of organization is organizationally closed. However, in processing structural inputs—energy and nutrients—it is structurally open. Autonomy and independence can be differentiated: systems may be *autonomous*—in that they produce their own organization—yet *dependent*—in that they require structural inputs from their environment.

Maturana and Varela conceptualized autopoietic systems with two basic attributes. First, they are characterized by a continuing process of production. The system has a pattern of organization that enables it to perpetuate itself by producing its own structural components from structural inputs and organizing the relations among these components in such a manner as to ensure the continuation of its own pattern of organization. Second, such a system produces its own boundaries as "surfaces of cleavage" that identify it as a composite unity separate from its environment.

I refer to the process of production as *poiesis* and—rather than take it as relevant only to *auto*poietic systems—I apply it to a more general class of systems that exhibit this continuing process of production, which includes *sym*poietic systems.[7] Sympoietic systems differ from autopoietic systems, in that they are *organizationally ajar* and *boundaryless*. Further distinctions between the two systems are listed in Table 6.1. I illustrate by providing a sympoietic description of ecosystems.

Table 6.1 Comparison of Poietic System Characteristics

Autopoietic Systems	Sympoietic Systems
Defining Characteristics	
Self-produced boundaries	Lacking boundaries
Organizationally closed	Organizationally ajar
External structural coupling	Internal and external structural coupling
Characteristic Tendencies	
Autonomous units	Complex, amorphous entities
Central control	Distributed control
"Packaged," same information	Distributed, different information
Reproduction by copy	Amorphous reproduction
Evolution between systems	Evolution within system
Growth/development oriented	Evolutionary orientation
Homeostatic balance	Balance by dynamic tension
Steady state	Potentially dramatic, suprising change
Finite temporal trajectories	Potentially infinite temporal trajectories
Predictable	Unpredictable
Advantages/Disadvantages	
Efficient	Adaptable, flexible
Constrained, codified information	Open to new and different information
Require certainty	OK with suprise

We generally conceive of ecosystems as composed of interacting parts—organisms as well as geophysical elements such as soil and climatic factors—working together through a variety of processes to create a definable whole. Similar to organisms, the identity of any particular type of ecosystem arises from the composition of these parts and processes and their arrangement—the pattern of their organization—that we use to distinguish among forests, grasslands, and wetlands, for example. Each of these different patterns of organization can be manifest in different structures. An eastern mixed-wood forest is different from a West Coast temperate rainforest; a short-grass prairie, different from a tall-grass prairie. Unlike organisms, however, ecosystems do not have the same kind of genetic "program" determining their identity, nor do they have self-produced boundaries.

For example, although many would readily call the West Coast rainforest noted in the introduction a system, the boundaries are by no means apparent. Especially when standing within, the sense is overwhelmingly of components and relations. Even when flying over the forest, the only clearly discernible boundaries are shoreline and clear-cuts and, moving up the flanks of the taller mountains, the gradual transition into bare rock and snow. While these may be boundaries, they are not "self"-produced. For example, even the treeline is a result of biological as well as climatic factors. If the latter are included as part of the system producing the boundary, then the boundaries must be drawn to incorporate these components, which would include a greater spatial extent, moving the boundaries further out.

Delineating boundaries by choosing particular criteria to distinguish in and out, is one means for conceptualizing such a system. However, "In principle an observer has the freedom to choose the system boundaries, he can decide which objects in the world he wants to take together. Of course he cannot decide whether his choice will be successful, this is a matter of experience" (Dalenoort 1989, 301). I suggest that an alternative, especially in those cases where the choices are problematic, is to relinquish boundaries. As noted earlier, the overwhelming impression in the forest is relation and interaction. Rather than produce boundaries to separate them from environments, ecosystems seamlessly integrate with their environments—and do so to the extent that the need to relate the concept of "ecosystems" to a three-dimensional physical space makes the abstraction problematic. Kimmins (1987, 27), for example, considers using the notion of "biogeocoenose"—a place-based concept—instead of "ecosystem," to avoid reification.[8]

Admitting ecosystems as boundaryless, however, does not equate to the disquiet associated with "anything goes." Rather, ecosystems seem to offer a fine example of reality resisting: there appear to be patterns and identities—even if not definable boundaries—that require some degree of negotiation for conceptualization. The phenomena that are typically described as "ecosystems" are conceived as having recognizable patterns of organization and structures and to limit changes in such patterns and structures in a self-determined manner. I refer to such systems, as organizationally ajar—as systems that allow, but restrict, changes to their patterns of organization. Rather than refuse inputs to prevent change in their pattern of organization, for example, ecosystems allow the introduction of new species, which may alter their identity considerably. However, ecosystems are not completely open, since inputs must be able to structurally couple in order to integrate into the system. For example, a plant species with a structure that cannot tolerate wet and shady conditions will not be integrated into a West Coast rainforest. A species that does have a structure able to successfully couple will be incorporated and has the potential to alter the forest's pattern of organization—even to the degree that a forest becomes a grassland, or vice versa.

These descriptions indicate—as noted earlier—continual challenge in considering "systems"—including autopoietic and sympoietic systems—as heuristics rather than as existing entities. For example, I find it simple and straightforward to interpret organisms as autopoietic and ecosystems as sympoietic, but the reverse seems more difficult. The temptation to think that organisms *are* autopoietic and ecosystems *are* sympoietic is tantalizing. While this could illustrate reality resisting and establishing the need for negotiation, I think it more likely reflects that I and others have been educated/socialized to interpret most complex systems, including ecosystems, as having boundaries and therefore as autopoietic. For example, education/enculturation in the fairly common "harvesting" mentality at the University of British Columbia's Faculty of Forestry—the incubator for British Columbia's forest practice for many years and where I completed my undergraduate degree—promotes interpretation of forest ecosystems as autopoietic. Silvicultural techniques are designed to enhance the "life cycle" of a forest by shortening developmental stages through planting and weed treatment. Stand-density diagrams are used to plot expected growth, with the intention of planning various treatments (spacing, pruning, fertilization), across the 80–130 years required to create a marketable crop.

Dissatisfaction with the discrepancy between my experience and this autopoietic conceptualization led me to form the concept of sympoiesis. Further refinement of the concept has facilitated interpretation of ecosystems in a different manner, as described earlier. To enrich both description and subsequent understanding, it is instructive to reflect on metaphors encouraging similar shifts in thinking. I find postmodern/poststructural contemplation fruitful in this regard, particularly Deleuze and Guattari's metaphorical use of "rhizome"—a decidedly sympoietic concept. To illustrate, I quote at length:

> In contrast to centred (even polycentric) systems with hierarchic modes of communication and pre-established paths, the RHIZOME is an accentred, non-hierarchical, non-signifying system without a general and without an organizing memory or central automaton, defined solely by a circulation of states....[A rhizome] is not made of units but of dimensions, or rather of shifting directions. It has neither beginning nor end, but always a middle, through which it pushes and overflows....Such a multiplicity does not vary its dimensions without changing its own nature and metamorphosing itself. Unlike a structure defined by a set of points and positions, the rhizome is made only of lines: lines of segmentation and stratification as dimensions, but also lines of flight or of deterritorialization as the maximal dimension according to which, by following it, the multiplicity changes its nature and metamorphoses. (Deleuze and Guattari 1983, 47–48)
>
> Where a tree is a single vector aimed at a specific goal, the rhizome expands endlessly in any number of directions, without a centre.... Its multiplicity is part of its nature, not its by-product. (Mansfield 2000, 142–143)

In drawing attention to Deleuze and Guattari's differentiation between trees and rhizomes, Mansfield points to reasons that echo my own decision to continue developing the notion of sympoiesis: "As with all metaphors that have become uncontested and obvious, the reasons for this usage [of tree as a predominant metaphor] are often seen as simple. Things grow and diversify the way trees do, we believe. But what assumptions and investments are preserved uninterrogated in this sort of metaphor?" (Mansfield 2000,141). This is essentially the same question that I previously raised regarding unreflexive use of "boundaries." Further, regarding the development of an alternative concept:

> This is no extravagant repudiation of the truth of the biological functioning of structures like the human body. Instead, what is being challenged is the simple assumption that things are to be understood as autonomous and separate, holding their truth in their coordinated internal structure. Rhizomatics rejects the idea that we can ever arrive resolutely at the advanced separation of things from one another which is the minimum starting point for the traditional representation of the world as the collocation of autonomous units. (Mansfield 2000, 147)

In the previous discussion, it was quite simple to describe organisms as autopoietic and ecosystems as sympoietic. Rather than claim the reverse as inappropriate or irrelevant—in other words, that the sympoietic heuristic should not be applied to organisms—I wonder what insights might emerge from such application—even while finding it a challenge. As an example of such a possibility, Deleuze and Guattari describe the physical tracing of orchid to wasp and wasp to orchid, each shape molded to the other, with a subsequent mutual reliance for continued poiesis: "wasp and orchid thus make a rhizome" (Deleuze and Guattari 1983, 19). This conceptualization of the wasp–orchid system—as well as any other symbiotic relationship—expands the notion of organism into something more sympoietic.

Another example is to question the degree of organizational closure in organisms that have been genetically modified or domesticated or hybridized. Does the pattern of organization in such cases include the genetic technician, farmer, or horticulturalist? What about organisms that have been fertilized or treated with pesticides or biological controls? A sympoietic interpretation of long straight cucumbers and large bananas would include social preferences as part of the system.

Regarding the identification and description of boundaryless sympoietic systems, I make one final but critical point. If a person considers a set of phenomena as a sympoietic system—and then draws a boundary around the phenomena/system for convenience or to establish a "working definition"—*that individual is no longer interpreting the phenomena as a sympoietic system.* Conceptualizing something as sympoietic requires relinquishing boundaries. Conceptualizing the same set of phenomena *with* boundaries would make it a different kind of system. Boundarylessness cannot be described with recourse to boundary delineation as a means of making conceptualization more convenient or simpler.

This raises the question, of course, about how we might identify or define a system, or speak of any set of phenomena as a whole or an entity, without delineating boundaries. As I have noted, however, we do this unwittingly to a substantial degree already. We speak of many different types of biological, social, cultural, economic, and other "systems" without reference to their boundaries or without clear definition of their components or extent. In more methodical or specific contexts, more methodical approaches may be required. Discussion of some possibilities for description and delineation can be found elsewhere (e.g., Dempster 1998b, 1999, 2003, 2004).

Closing Comment

The foregoing discussion has set out a theoretical platform—based on systems theory—for conceptualizing complex situations and phenomena as boundaryless. Brief descriptions and distinctions between organisms and ecosystems—based on an ecological understanding—have exemplified the possibilities for application. Discussing the myriad implications that arise is well beyond the scope of this chapter, which has primarily focused on the conceptualization of sympoietic systems as boundaryless.

As noted earlier, phenomena can be distinguished by drawing boundaries, by drawing comparisons, or by drawing relationships. Each approach offers benefit and disbenefit. In this chapter, the emphasis has been on *relinquishing* the need for drawing boundaries by focusing on drawing comparisons—in particular, between the characterizations of autopoietic and sympoietic systems. Brief reference to Deleuze and Guattari's discussion of rhizomes also draws away from the need for boundary delineation through application of a boundaryless metaphor.

Similarities and differences in the defining attributes of autopoietic and sympoietic systems lead to differences in characteristics and behaviors between the two system types—including the different advantages and disadvantages—that were listed in Table 6.1. Perhaps the most beneficial consequence of conceptualizing the distinction between the two system types is the ability to provide a theoretical basis for grouping each set of characteristics. Although any set of phenomena interpreted through these heuristics will not unconditionally fit either set of characteristics, an interpretation that mixes "opposite" characteristics (e.g., centralized control and adaptive) would raise questions regarding the appropriateness of the interpretation. In this way, these heuristics can be applied to grapple with understanding of many complex systems—such as the forest ecosystem and its broader connections or relations to human economic, social, cultural, and educational systems that was described at the outset.

Notes

1. I acknowledge discussion with my colleague Eric Tucs on these points.

2. I should note, however, that such a stance is not illustrative of Wilber's work. Rather, he emphasizes the value of "no-boundary thinking" in a very no-boundary thinking kind of way.

3. Autopoiesis has also been applied to cognition and has contributed to radical constructivism (e.g., von Glasersfeld 2001) and conceptions of embodied knowledge (e.g., Maturana and Varela, 1987, Varela et al. 1991). These are valuable epistemological interpretations and theoretical directions—which I will not endeavor to cover here—although they have influenced my onto-epistemological position described at the outset.

4. Maturana and Varela used the term "organization" to describe this concept. I follow Capra (1996) to avoid confusion with human organizations such as institutions.

5. Note that this definition fits a vernacular understanding of structure, which typically refers to a physical entity—something present and "real"; it does not match

the definition applied in some disciplines. For example, in some cases, such as when contrasted to process, structure more closely represents what is here being termed pattern of organization.

6. Maturana and Varela used the notions of open and closed—*ajar* is my own addition to recognize the different process in sympoietic systems.

7. Note that poiesis, refers to "*self*"-*production*—continual production of the *same* system—not *reproduction*, which refers to production of *another* system, another structure with the same pattern of organization.

8. This may, however, be an example of externalizing, rather than removing or addressing, the problem.

References

Bird, Elizabeth Ann R. 1987. "The Social Construction of Nature: Theoretical Approaches to the History of Environmental Problems." *Environmental Review* 11(4): 255–264.

Bunch, Martin. 2001. e-mail communication.

Capra, Fritjof. 1996. *The Web of Life: A New Scientific Understanding of Living Systems*. Garden City, NY: Anchor Books.

Chadwick, George F. 1966. "A Systems View of Planning." *Journal of the Town Planning Institute* 52(5): 184–186.

Dalenoort, G. J., ed. 1989. *The Paradigm of Self-Organization: Current Trends in Self-Organization*. New York: Gordon and Breach.

Dean, Jodi. 1996. *Solidarity of Strangers: Feminism after Identity Politics*. Berkeley: University of California Press.

Deleuze G., and F. Guattari. 1983. *On the Line*. Translated by J. Johnson. New York: Semiotext(e).

Dempster, Beth. 1995. *System Stability and Implications for Sustainability*. Unpublished B.Sc. Thesis, University of British Columbia, Vancounver, B.C.

———. 1998a. *A Self-Organizing Systems Perspective on Planning for Sustainability*. M.E.S. Thesis, University of Waterloo, School of Urban and Regional Planning. <www.bethd.ca/pubs/>

———. 1998b. *Conceptualizing Complex Systems: A Methodology for Characterizing Systems Relevant to the Planning and Management of Parks and Protected Areas.* Technical paper #12. Heritage Resources Centre, University of Waterloo, Waterloo, Ontario.

———. 1999. "Post-normal Science: Considerations from a Poietic Systems Perspective." Unpublished hypertext. <www.bethd.ca/pubs/>

———. 2000. "Sympoietic and Autopoietic Systems: A New Distinction for Self-Organizing Systems." In *Proceedings of the World Congress of the Systems Sciences and ISSS 2000.* J. K. Allen and J. Wilby, eds. Presented at the International Society for Systems Studies Annual Conference, Toronto, Canada, July 2000. <www.bethd.ca/pubs/>

———. 2004. "Canadian Biosphere Reserves: Idealizations and Realizations." *Environments* 32(3): 93–99.

———. 2003-2006. *sympoiesis.net* <www.sympoiesis.net>

Kimmins, J. P. 1987. *Forest Ecology.* New York: Macmillan.

Mansfield, Nick. 2000. *Subjectivity: Theories of the Self from Freud to Haraway.* New York: New York University Press.

Maturana, H. R. 1980a. "Man and Society." *Autopoiesis, Communication and Society.* F. Benseler, ed. Frankfort: Campus.

———. 1980b. "Autopoiesis: Reproduction, Heredity and Evolution." *Autopiesis, Dissipative Structures, and Spontaneous Orders.* M. Zeleny, ed. Boulder, CO: Westview Press.

Maturana, H. R., and F. Varela. 1980. *Autopoiesis and Cognition: The Realization of the Living.* Dordrecht: D. Reidel.

———. 1987. *The Tree of Knowledge.* Boston: Shambhala.

Varela, F., H. Maturana, and R. Uribe. 1974. "Autopoiesis: The Organization of Living Systems, Its Characterization and a Model." *Biosystems* 5(94): 187–196.

Varela, F., E. Thompson, and E. Rosch. 1991. *The Embodied Mind.* Cambridge: MIT Press.

von Glasersfeld, Ernst. 2001. "Distinguishing the Observer: An Attempt at Interpreting Maturana." www.oikos.org/vonobserv.htm [January 27, 2001].

Weinberg, Gerald M. 1975. *An Introduction to General Systems Thinking.* New York: John Wiley and Sons.

Wilber, Ken. 1985. *No Boundary: Eastern and Western Approaches to Personal Growth.* Boston: Shambhala.

Part II
Community, Values, and Sustainability

Chapter 7

Boundaries on the Edge

Irene J. Klaver

> But at what moment does wood become stone, peat become coal, limestone become marble? The gradual instant.
> —Anne Michaels, *Fugitive Pieces*

Intermediate Cases

Old Faithful, Yellowstone's most famous geyser, got its name because of the regularity of its eruptions—every hour and a half it spouts four to eight thousand gallons (14,000–32,000 liters) of boiling water 130 feet up in the air. The name Old Faithful pertains as much to the water itself as to its regular eruptions. The water is old and has stayed faithful to a long journey. For 500 years it percolated its way down and up again, cooled and heated over the long years to erupt for only one to four minutes—a culmination of the slow and the sudden. Anne Michaels calls this kind of moment the "gradual instant," which perfectly captures its oxymoronic nature.[1] A long duration works together with the instantaneous toward this definite, but undecidable, boundary moment when water turns into a geyser. In other instances of the gradual instant: wood becomes stone; peat becomes coal; and limestone marble. These moments are nothing mysterious, they are all around us—it is the way grass greens, people get angry or fall in love, revolutions take place. We are surprised when it happens, but often know already that the radical change was a process in the making.

In order to understand how processes of change work, an understanding of the structure and function of boundaries is crucial. They are places not only of demarcation and separation but also of potential transition, transformation, and translation. The prefix "trans-" in all these cases denotes the possibility of crossing, of going over, beyond, across. Boundaries can be crossed. In this chapter I explore what I call the edge of boundaries, that is, their transitional dynamic or force. How do transitions take place? How can they be facilitated, prevented, or not disturbed (so as to protect various natural processes)? How do boundaries themselves change in the course of transition?

Understanding, according to Wittgenstein, "consists in the very fact that we 'see connections.'"[2] This is a transformative process—it changes our interpretative frameworks. Boundaries are intrinsically part of this dynamic; either found or constructed, they are the places or moments where things are exposed to each other. Boundaries separate domains, and by this very act create the domains that they separate. They engender new configurations and patterns, which elicit new understanding—which, in turn, leads to new boundaries. This is a reciprocal and reiterative process, in which understanding and boundaries repeatedly redefine each other. Understanding, by Wittgenstein's definition as the seeing of connections, is a boundary crossing and any such transition entails a new boundary drawing. Understanding can thus be characterized as a shifting of boundaries.

To see how things—be they ideas, ecosystems, worlds, or words—are connected is not always self-evident. Therefore Wittgenstein adds emphatically: "Hence the importance of finding and inventing *intermediate cases*." For example, only when we realize how the widely used fertilizers of America's intensive agriculture drain into groundwater, how a network of water veins enters into tributaries accumulating and transporting nitrogen in the mighty Mississippi, do we begin to understand that the dead zone[3] in the Gulf of Mexico is connected to land management thousands of miles away in the Midwest. This understanding reveals, furthermore, that the boundaries of a river are not only formed by its banks or even floodplains, but as much by its watershed or basin—the whole area of land that drains into that particular river. Only when one can make these connections—from what I call a water basin-mentality—can one understand that good water management is intrinsically related to good land management.

It is interesting that Wittgenstein gives equal importance to both the *finding* and the *inventing* of these intermediate steps. No realism-constructivism issue concerns Wittgenstein. What counts for him, is facilitating the versatility of an inquisitive and creative mind that leads to understanding. It doesn't matter if intermediate cases already exist and await being found, discovered, or searched for; or if they are (hu)man-made, constructed, invented, or created. Both approaches are fine with Wittgenstein, as long as they generate new understanding. And, in fact, finding and inventing can spark each other, work together in a continuous process of translation and re-interpretation, mediating each other. Bruno Latour speaks of a circulating reference in this context.[4]

That does not mean, however, that finding and inventing are the same. Think, for example, about the way we conceptualize the history of the West.

Saying that Lewis and Clark discovered the West seems very different from saying that they made or invented the West. But suppose we say, more ambiguously and less straightforwardly, that with them the West was engendered. After all, through their expeditions (their discoveries) and their writings (their in(ter)ventions) an understanding of the West, came into existence—a narrative of a land stunningly beautiful, wild, and free. Both renderings—finding, inventing—make different histories visible, set different trains of thought or agendas of action in motion. Depending on what one wants to accomplish—in this case naturalizing and reifying the myth of manifest destiny or debunking it—an emphasis on finding or on inventing will be most pertinent. Again, what counts is the connections that are made and the understanding that is generated.

In the rest of this chapter I explore the transitional and transformational capacities of boundaries—the workings of boundaries on the edge—in part by analyzing various ways that intermediate cases can be found or invented to generate understanding. I trace how boundaries can facilitate cultural, social-political, ecological, and epistemological transitions, especially through the workings of boundary objects. More than once, I invoke the power of oxymoronic structures to overcome exclusive, dualistic boundary determinations. I transverse ecotones, edge effects, watersheds, corridors, green eyes, the Berlin wall, coyotes, and bird migrations to finally come to the edge of the grass where a new boundary awaits. I begin with boundary objects and an analysis of how and why they work by examining a most forceful case: Don't Mess with Texas.

Boundary Objects—Don't Mess with Texas

"Don't Mess with Texas" is the cleverly constructed slogan with which the Texas Department of Transportation (TXDOT) kicked off its 1986 tough-talking litter prevention campaign. Meant to educate all Texans about the litter problem in their Lone Star State, it was featured widely on television, radio, and billboard, with local celebrities, such as singer-songwriter Willie Nelson, promulgating the cause. For a campaign telling people to keep their trash in the car and off the road—no more burrito wrappers tossed and beer bottles hurled from the pickup—it became widely popular. It turned into its own product line—caps, mugs, T-shirts, key chains, bumper stickers, you name it.

The campaign was a huge success. Ninety-six percent of Texans came to know the slogan, and litter on Texas roadways has been reduced by fifty-two percent since 1995. The slogan managed to capture the spirit of Texans themselves: a tough, independent, don't-tell-me-what-to-do kind of folk.

The strength/success of the slogan, Don't Mess with Texas, is that it connects different worlds or frames of mind. It speaks to the mentality of the independent cowboy as well as to the world of environmental concern—both appealing to a sense of belonging to Texas. Thus, it forms a perfect boundary object between worlds that have little in common, inhabited by people who often despise each other. The crux, the bridging element, is the double meaning/function of the word "mess." It facilitates transition, the moment and place where the two worlds can meet and work together, and crystallize into a circumstance, a technique of connecting—a concrete connection in a particular thing and activity. Don't tell a Texan what to do; don't mess with him; mind your own business. But, also, don't touch his Lone Star State; don't mess with it; don't make a mess of it; keep it clean, keep it beautiful. In both cases, there is a convincing appeal to Texas pride—different sides of pride. The don't-give-a-shit roustabout or ranch hand and the concerned environmentalist become unexpected bedfellows through a playfully ambiguous use of the word "mess."

A boundary object such as Don't Mess with Texas functions as an intermediate between heterogeneous groups. Star and Griesemeier developed the term "boundary object" in the context of scientific fieldwork where diverse groups try to achieve common understanding or collaboration across disciplinary divides through "translation of each other's perspectives."[5] The same applies to many other situations where people from widely different backgrounds have to find common ground.

Sustainable management, for example, depends on successful, ongoing negotiation of often mutually exclusive interests.[6] The art of this negotiation is the ability to reach practical, communally crafted decisions in which various concerns are taken seriously, so that concessions can be made. To understand each other's concerns, they must be "translated" into one another's conceptual and practical vocabulary or framework. Boundary objects facilitate this translation: they have a specific meaning for each group, but their structure is common enough to form a leverage or connection among the various groups. They can be seen as vehicles of translation that enable coherence and understanding across social worlds; in their capacity to bridge boundaries, boundary objects create a meeting ground, a connection. They invite cooperation, without reducing diversity, to reach a consensus that allows for heterogeneity. In carefully developing this cooperation, vested interests might shift, attracting potentially new players to the field, eliciting new understandings. Cooperative management is, by definition, adaptive management because it is based on ongoing processes of negotiation.[7]

To use the same word with a different slant, or twist—epitomized by Don't Mess with Texas—is one secret to bridging a potentially divisive structure. Next, I explore another.

Spatial and Temporal Transitions

Boundaries are of various kinds—spatial, temporal, and qualitative. "They are different as night and day," we say, because nothing is more different from the dark of the middle of the night than the brightness of mid-day. But at the transition periods of dusk and dawn edges fade—dark and light turn gradually into each other, fluidly, almost imperceptibly. Lacking a distinctive moment of transition, they are transition periods, gentle corridors of time, carrying day and night smoothly into each other, helping the world wake up and fall asleep by softening the boundaries between night and day. These temporal corridors create special events. Birds start to sing at dawn and plants open up, while deer bound and owls hover at dusk. Similarly, spring and fall accommodate transitions between summer and winter. They are corridors of time facilitating/affording certain activities: major migrations are set into motion, and trees lose their leaves in preparation for the cold or leaf out in preparation for growth and reproduction. As natural beings, we, too, are temporal beings; time counts for us as well—night/day, dawn/dusk, the shifting seasons. And just like the water under Old Faithful, things build up... and then we get angry or sad or happy and glad.

Dusky transition zones are often needed to facilitate processes of regeneration. Once upon a time in Oregon, for example, there stood an old growth Douglas fir forest. Then many were clear-cut. The best management practices of forestry prescribe that they be restored. On the open patches of land left alone, manzanita moved in first. To hasten the restoration process, the U.S. Forest Service planted young firs in the clear-cut soil before the manzanita got a chance to establish. All firs died. It turned out that the manzanita produces a mycorrhizal microbial community that the soil needs in order to be ready to "receive" the firs. To facilitate and enable the sharp qualitative boundary between bare soil and mature Douglas fir forest, we need an interlude, a temporal boundary of intermediate breadth—manzanita.

In mature Douglas fir forests, spotted owls need the areas where they do not nest as thick boundaries between breeding territories. The manzanita constitute a temporal boundary, an intermediate successional sere from bare ground to mature Douglas fir forest. The owls need a similarly broad spatial boundary of owl-less woods to demarcate their territories. The pattern, abstractly, is the same: day and night, but also—and just as important and necessary—dawn and dusk; summer and winter, but also—and just as important and necessary—spring and fall.

Most things don't have hard, well-defined, clear-cut, or stable boundaries. Most are like mold, spreading in unpredictable patterns. Mold's only hard boundary is determined by a crucial moisture gradient, but its spatiality is hard to fathom. Could one ever predict which shape it will take? Producing tiny spores to reproduce, they waft imperceptibly in and out of our existence, in and out of our homes, our lungs. Wherever enough moisture and an appropriate absorbent surface interact, the spores take hold and grow, living off the material they landed on, to become mold.

Manzanita, mold, owl-less Old Growth, dawn/dusk, spring/fall, everything is a means for something else, no end to transition, no isolation—thus, no ends in themselves, no intrinsic value, no possibility of thinking beyond means. Thinking is refocused on a path, a transition zone, a temporal corridor as a mean (and means) mediating transitions. To create connections between various entities, whether political groups or scientific realms, as well as between pieces of wood or fiberglass, one needs a common linkage, intermediate cases, boundary objects. In the case of systems of thought, this linkage comes in the form of conceptualizations. For example, in ecology, the "selection of a specific boundary for investigation, ultimately depends on the research question."[8] This partly arbitrary boundary of an ecological "isolate" will consequently generate new questions and thus new boundaries.

Boundaries can be agents of change, corridors of communication, and can thus lead to transitions, or elucidate transitions. When they are seen as final, definitive, they are confining—that is, *con finis* (Latin for "with end")—they entail an ending and have lost their generative connection to beginning. The pragmatic question is how do boundaries turn into places of transition, places where things begin and end, and ends become means? How and when are they places of transition instead of closure? When are they impervious to transformation and translation? How can certain processes be facilitated and others be prevented?

In the next section, I suggest how juxtaposition of opposites can function as a facilitating process of translation, generating new meanings at the boundary lodged in the concept of an oxymoron.

Oxymorons

The word "oxymoron" is neat. Being a paradoxical concept, it *is* what it means—derived from the Greek *oxy* ("sharp") and *moros* ("dull"). An oxymoron is a figure of speech that combines two contradictory terms. Of course in the word "oxymoron," the two opposing terms are within the word, while usually the oppositional boundary of an oxymoron is created by two words that mean opposite things, but, in their juxtaposition, gain a special meaning or force. "Deafening silence" is a perfect example. There is a sharp boundary between the two concepts, like in the "gradual instant," but in their contiguity they acquire a new meaning, a transitory closeness, a belonging together.

Some oxymorons are simply contradictions in terms—like violent nonviolence. Some are droll and mildly scornful—like "sophomore," literally a wise fool. Some juxtapositions are declared to be oxymoronic in order to mock or deride—like an honest lawyer. Some enable a deep psychological insight—like passive aggressive. Oxymorons in general are effective because they place seeming opposites side by side. In this juxtaposition they challenge traditional dualisms that dichotomously split our cultural imagination into two distinct and separate realms. Dualisms have been an important feature in a long tradition in Western culture. We are still prone to classify things in terms of these clear-cut dichotomies. All thought, Plato avers in the *Timaeus* and his other "later" dialogues, is a matter of the same and the different. Other dualisms besides same/other include: culture/nature, mind/body, reason/emotion, man/woman, good/bad, light/dark, white/black.

Dualisms are not innocent. The terms at either side of the slash are valued differently—the category of the other systematically undervalued in favor of the category of the same, or, culture favored over nature, mind over body, reason over emotion, man over woman, good over bad, light over dark, white over black, and so on. Thus an implicit system of interlocking oppressions is formed that seems to legitimate (if not necessitate) an exploitation by one side of the slash over the other.[9] The beauty of oxymorons is that they put both sides together and thus bring out the complementarity—or what I call co-constitution—of the concepts instead of their exclusion. They are mutually constitutive instead of mutually exclusive. You can't have one without the other. Oxymoron opens the dualistic structure to a continuum: white and black are the extremes of a continuous scale.

Our dualistic habits of thought are important to keep in mind when we try to negotiate conflict or invoke new ways of thinking. They require creative and imaginative approaches—sometimes a complete reversal of a specific mode of operating. The central (and revolutionary) tenet of Aldo Leopold's philosophy of game management invokes precisely this emphasis on creativity: "game can be restored by the *creative use* of the same tools which have heretofore destroyed it—axe, plow, cow, fire, and gun." The task of the game manager is to find and apply the creative instead of destructive use of these tools.[10]

A philosophy that takes nature—the world of matter—seriously has to understand how co-constitution works and can no longer allow for a system of interlocking exclusions that legitimates exploitation of natural entities. Oxymorons have been effectively employed in contemporary philosophy to overcome a dualistic tradition. Ann Michaels, as noted, uses the term "gradual instant" to evoke the dual temporalities simultaneously at work in the "moment" when peat turns into coal. Cultural anthropologist Arjun Appadurai captures a similar oxymoronic or paradoxical moment with the term "vernacular globalization."[11] Where these two concepts are usually mutually exclusive he foregrounds how they can work in a mutually constitutive fashion, that is, how global processes *create* as well as displace regional distinctiveness. Local cultures have been obliterated by global economies, but have also revitalized themselves by incorporating international cultural elements. Local identities have disappeared but also reinterpreted themselves in terms of cultural globalization—including entrenching themselves in violently radically conservative identities. Interestingly, the term "radically conservative" itself would in the 1960s have been an oxymoron, but has lost its oxymoronic status when one considers radical Christian or Islamic fundamentalism.

Michel Foucault sees the oxymoronic dynamic as the center of philosophical discourse: philosophy should develop an "ontology of the present" by determining an element of the present that elucidates "what this present is."[12] Ontology of the present carries the same paradox as Arjun Appadurai's vernacular globalization. Where "ontology" refers to "supra-historical structures of being," the "present" refers to the particular event of a "here and now." Juxtaposing the two concepts into one term, Foucault draws the incompatibility of "ontology" and "present" into an oxymoronic tool that brings both concepts into play for understanding any particular phenomenon.

Jean-Luc Nancy makes a similar move in describing the world as "the infinite resolution of the finite," as does Judith Butler when she speaks of an "ongoing cultural articulation of universality."[13] All try to grasp how these opposite movements arise in one and the same moment, continuously shaping and readjusting each other's borders in a constitutive tension. Merleau-Ponty called these moments of opposite movement "sedimentation and reactivation" and determined their relationship to be the most "fundamental problem."[14] An "ontology of the present" always brings "its" ontology into reactivation, "its" present into sedimentation—in one and the same movement and moment.[15]

The symphonic workings of oxymoron in the conceptual domain have an ecological equivalent in the workings of ecotones.

Ecotones

The boundary place where different natural habitats meet is called an ecotone. An ecotone is an interface between two ecosystems and is often a more complex ecosystem, with its own processes and species. Because it is a biological crossroad, one easily finds a great variety of life forms in these transition areas—some of them unique to the ecotone, found in neither of the two meeting ecosystems. Where hardwood forest meets long grass prairie, one finds Aldo Leopold's beloved "oak openings"—a savannah containing forest species, prairie species, and species all of its own. Where boundaries are sharper, as are those between pasture and woodlot, or sea and land, edge effects are the boundary workings of an ecotone; they take place at the interface or boundary between the two ecosystems. Edge effects create a shared boundary space with its own diverse and dynamic activities. Thus, at the shoreline we find a rich ecosystem full of crabs, clams, fish, along with the place where turtles lay their eggs. When the tide recedes, herring gulls pick their food between the exposed rocks and, screaming high in the sky, drop the clams on the very same rocks to break their shells.

Rachel Carson captures the edge effect of the shoreline—the ecotone between sea and land—on a geological temporal scale:

> Once this rocky coast beneath me was a plain of sand; then the sea rose and found a new shore line. And again in some shadowy future the surf will have ground these rocks to sand and will have returned the coast to its earlier state. And so in my mind's eye these coastal forms merge and blend in a shifting, kaleidoscopic pattern in which there is no finality, no ultimate and fixed reality—earth becoming fluid as the sea itself.[16]

The rising sea finds a shoreline, which is partly created or *founded*, by the sea itself. The rolling waves of the sea shape a place, the shoreline, that reciprocally affects the ensuing movements and reach of the sea. The finding of the shoreline is intrinsically related, intertwined, with its own founding; the finding and founding co-constitute each other in one and the same act. Where the finding has a constitutive aspect, the founding has a discovering aspect. Wittgenstein's finding and inventing cannot be reduced to the same thing, nor are they simply interchangeable; rather they are different (em)phases of the same process. Thus, in a way to say that the sea founds (both finds and invents) the shore is oxymoronic, but no more so than to say that the shore is porously impervious. The difference is important insofar as each emphasis leads to another trajectory of exploration, another possible practice. Did Lewis and Clark invent or discover the West? Just as the ocean and the shore shape each other, we can say that the western landscape (including its indigenous inhabitants) and Lewis and Clark's writings shaped each other. They mutually constituted each other. The particularities of the landscape drew out the writing, the writing framed the landscape in terms of a nineteenth-century sublime discourse, and the West as we still conceive of it today was born in our cultural imagination.

When an ecologist poses questions about the ecosystemic effect of, say, beavers in Yellowstone's Lamar River valley, he or she implicitly bounds an ecosystem—the Lamar River watershed or water basin. When another ecologist poses a question about the ecosystemic effects of wolves in the

Yellowstone, another, larger (at least) park-wide ecosystem is bounded—one that contains the Lamar valley. Does the latter cease to be found as a whole and entire entity?

To start with the sea finding the shore, the focus is on the already existing lay of the land—the rhythms of its constitution are of another temporal scale (see Callicott's chapter in this volume) than the ones of the tides. To focus on the sea founding/creating the shore is to analyze the way the waves mold the shore they encounter. On a temporal scale of the everyday there might be stability, but on a geological scale there is fluidity. With the ongoing rhythm of ebb and flow, the sea grinds stone to sand and coastal forms merge and blend. Everything is active: "the sea rose and found a new shore line." Again: this finding has a founding/constitutive aspect to it. The sea founds/constitutes a new shore by means of her own presence—by rising, the sea enters the land and finds the rocks of a new coast lying at the feet of her waves. Finding and founding come together in this sense of "found." Estuaries are another interesting sea/shore ecotone in which dynamic ecosystems form buffer zones between saline seas or oceans and a river's freshwater. Savannahs, shorelines, estuaries and other ecotones are nature's founding boundaries. Next I find some generated by and generative of culture.

Boundary Projects: Recoding the World

Nature and culture instigate cascades of interwoven movements and thus reveal their boundary as porous. Nature and culture have been treated as separate domains, a forceful dualism, completely embedded in the interlocking oppression of nature as being inferior to culture. But, as with every dualism, they not only operate together, but co-constitute each other. They are not one and the same, but their separation has often been artificially enlarged. In fact, the very boundary between nature and culture might itself be an artifact, made to accommodate an easy access to whatever was destined to be natural and hence legitimizing an overuse with often disastrous consequences for the exploited ecosystems.

A boundary is not a static entity, but operates, is at work, between various entities, mutually constructing them and itself in the working. Heidegger states in "Building Dwelling Thinking" that a "boundary is not that at which something stops but, as the Greeks recognized, the boundary is that from which something *begins its presencing*."[17] We could add that boundaries themselves begin their own "presencing" in this. That is, they themselves *materialize* in interactions. Furthermore, as *practices* they are—as here explained —generative: they instigate new directions, turn moments into events. They are also mediators: they trigger further developments, new trains of thought, new events. Boundaries are places at work. They are places in and of themselves, with their own causes and effects.

According to Donna Haraway, any boundary is in fact a "boundary project," a "site of production," that recodes the world—while the world recodes us—and in this process we need to revision "the world as a coding trickster with whom we must learn to converse."[18]

Learning to converse happens through translation, which is facilitated by boundary objects, intermediate cases, oxymorons, hybrids, and boundary projects. Boundaries are active. They are places of negotiation and translation; they change and induce changes. Therefore, it is important to understand their dynamic—important to ask how boundaries arise, how they function, how they can be negotiated, and under what circumstances they change. Finding and inventing intermediate cases is a question of retracing boundaries, seeing new patterns, finding our place in the world Haraway calls a "coding trickster."

The most exemplary "coding trickster" is the Native American coyote—the New World fox of European fairy tales. Fox and coyote permanently tinker with our boundaries. They follow us, but don't let themselves be domesticated; instead they eat our chickens, rummage through our garbage, and frustrate our efforts to do anything about it. They use us, within our own boundaries. The tricksters are the ones in between. They take many shapes and forms, not only in mythology but also in contemporary literature. For example, they look at us with the green eyes of a half-breed in Leslie Silko's novel.

Green Eyes

Aldo Leopold famously shot a wolf, a larger cousin of the coyote, and watched a "fierce green fire" die in her eyes. Tayo, the main character in Leslie Silko's celebrated novel *Ceremony*, is a half-breed with hazel-green eyes.[19] His eyes, green as the Paluxi River, carried shame in them. They reminded the Laguna reservation that some of their women had spent nights in Gallup, New Mexico. Tayo was born there, on Gallup's north side, in Little Africa, where the blacks, Mexicans, and Indians lived in cardboard shacks with tin roofs next to the river and the dump. His mom dropped him off at the reservation and Uncle Josiah took him in—not daunted by the shame in those green eyes. Josiah's lover was the green-eyed cantina dancer. He had bought spotted Mexican cattle from the cantina dancer's Mexican cousin because "If it's going to be a drought these next few years, then we need some special breed" to ride it out.[20] These Mexican cattle were not like the white-faced Herefords; their dark eyes were as green as brown can be. They would not wait at the gate for water to be brought to them, dying of thirst. Josiah's cattle were long-legged "descendants of generations of desert cattle, born in dry sand and scrubby mesquite, where they hunted water the way desert antelope did."[21]

When Tayo was in Vietnam, fighting in the war, Uncle Josiah died, and without any regard for the white man's fences, his long-legged cattle had moved southward, violating property lines along the way. The quest to recover the cattle became Tayo's ceremony to overcome his own boundaries and those separating his Indian and (Mexican-)American cultures. The Mexican cattle came to symbolize the lost knowledge of how to negotiate the closed boundaries of fixed fences and green eyes.

Tayo got his clues from Old Betonie, a medicine man whose hogan overlooked Little Africa. He had hazel eyes like Tayo himself. Mixed with his medicine man's traditional paraphernalia, such as skin pouches and deer-off clackers, Betonie had piles of telephone books, bundles of newspapers, and the Santa Fe Railroad calendars that were so common on the reservation, with Indian scenes painted on them. He kept them because "All these things have stories alive in them," as the old man said.[22]

> You see, in many ways, the ceremonies have always been changing.... At one time, the ceremonies as they had been performed were enough for the way the world was then. But after the white people came, elements began to shift; and it became necessary to create new ceremonies. I have made changes in the rituals. The people mistrust this greatly, but only this growth keeps the ceremonies strong.... Things which don't shift and grow are dead things.[23]

Betonie had shown him the privilege of being born between worlds, as well as the precariousness and the obligations that come with it, emphasizing that it is not an easy position. "It is a matter of transitions, you see; the changing, the becoming, must be cared for closely. You would do as much for the seedlings as they become plants in the field."[24] In time, Tayo's eyes became lush green fields, a shared space, where various worlds met and mixed. Green is the color of the field, of openness, of grass growing, the color of hope. The promise of the budding green in the spring is the annual recurrence of new life after a barren winter. Green as the symbol of hope is a natural referent in culture, a natural engendering of the cultural signification of hope. In the following I will invoke a cultural engendering of signification of hope through a ritual of Jane Goodall. She found in the stones that once were part of boundary/prison walls the symbol of hope.

Holes in the Flag, Holes in the Wall

When the Berlin wall came down in 1989, people took parts of the wall with them—pockets filled with stones, some sold later for high prices, although never as high as the price paid for the wall when it was still a solid barrier. The first hole—the not-yet determined—made in the wall symbolized hope: a promise of an open future, while at the same time a witness of a once imposing border. Here we find oxymoronic play again. Any opening testifies as much to the indeterminacy of a future as to the framing from which it appeared as opening; it derives its identity as opening and possibility only because it is framed in the wall that formed a pertinent closure. It testifies to an historical era because it is a hole in the Berlin wall, symbol of the cold war, symbol of a political constellation/era of two superpowers; the hole in the wall opening the door to new power relations/constellations that far exceed the boundaries of any unity. The hole made in the Berlin wall is like the hole in the flag of Romania. Slavoj Žižek explains:

> The most sublime image that emerged in the political upheavals of the last years... was undoubtedly the unique picture from the time of the violent overthrow of Ceauşcescu in Romania: the rebels waving the national flag with the red star, the Communist symbol, cut out, so that instead of the symbol standing for the organizing principle of the national life, there was nothing but a hole in its center. It is difficult to imagine a more salient index of the "open" character of a historical situation "in its becoming," as Kierkegaard would have put it, of that intermediate phase when the former Master-Signifier, although it had already lost the hegemonical power, has not yet been replaced by a new one.... What really matters is that the masses who poured into the streets of Bucharest "experienced" the situation as "open."... The enthusiasm which carried them was literally the enthusiasm over this hole, not yet hegemonized by any positive ideological project.[25]

Jane Goodall keeps a piece of the Berlin wall with her wherever she goes, along with a piece of limestone from the Robinson Island prison of South Africa. As ambassador for Gombe's chimpanzees and nature in general, she travels the world lightly—always, however, carrying these two stones with her. They are symbols for her mission: to counter apathy with hope. Where stones are usually a symbol for burden (cf. Sisyphus), she turns them into a symbol of hope. Her stones were once building blocks for walls to keep people separated; they are now transformed into references toward hope and connection. A larger solidity is found in the *fragments* of the walls. The wall and crushed regimes reveal the versatility of the solidity of stone: its materiality lends itself as easily to a symbolization of closure or oppression as to one of liberation.

When the Berlin wall came down in 1989, social-political and economic orders tumbled with it. Generations of Marxist intellectuals and walls of books lost their feeding grounds. Ecological conditions also changed. With the absence of military activity, the vast grasslands of Russian and Polish military bases quickly reforested, leaving flocks of migrating geese without foraging grounds. Both intellectuals and geese moved on to new fields. The academics took up themes such as the nation-state, multiculturalism, globalization, and the environment. The birds found new grazing grounds on their migration route, such as a large Dutch wetland area, the Oostvaarders Plassen, itself a result of the intricate relations between nature and culture.

Shifts in boundaries at one place have ramifications at other places—often unforeseen, sometimes unforeseeable, changing things in unexpected domains, at unexpected times. The unforeseen and unforeseeable constitute both the fear of bioengineering and the promise or hope of ecological restoration. These two interventions operate explicitly on the boundary between nature and culture; and most often this boundary is crossed unwittingly. Only through intermediate steps one can begin to understand how something like the fall of a wall can cause flocks of geese to descend on wetlands in the Netherlands and how geese, in their turn, have an effect on environmental policy. Natural and political events trigger chains of effects, crisscrossing each other's domains, thus testifying to the porosity of the nature–culture boundary, if not to the tentativeness of their very separation.

Just as dried soil needs to be opened up, to be broken, for the water to percolate, so our conceptual schemes need always to be opened for connections to be seen—or to be made—in order to understand. We need to create open situations in human sociopolitical settings. We need Tayo to find the value of his green eyes, just as Aldo Leopold eventually found value in those into which he looked. We need adaptive management strategies. We need to learn how to use our same tools differently. We need holes in the flags, in the walls, in our taxonomies, in our scrambled eggs.

Taxonomy of Boundaries and Boundaries of Taxonomies

Boundaries are multiple, multifarious, multifunctional, and multistructural. The conceptual boundary of the word "boundary" itself is unclear, transitional phases morphing into each other, its meanings spread out like mold, digesting the material they land on. There exists a mosaic of terms: there are boundaries, borders, borderlines, buffers, constraints, corridors, ecotones, edges, frames, fringes, frontiers, interfaces, limits, links, limina, passages, shorelines, shores, banks, transition zones, thresholds—each with its distinct right, with different connotations and functions, different roles to play. Any taxonomy will depend on practices involved, questions asked, lives lived—these will determine which patterns will appear as pertinent for any taxonomy of boundaries.

There are overarching themes that unite boundaries: boundaries are never static; they are always places of change, of negotiation. This makes it pertinent to understand their dynamic, how they arise, how they function, how they can be negotiated, under what circumstances they change, and what patterns can be discerned in their change. In the labyrinths of Borges's language, familiar categorizations lose their boundaries. Foucault's *The Order of Things* is born out of this collapse—out of laughter about Borges's animal taxonomy gathered from a "certain Chinese encyclopedia" that divides animals into: "(a) belonging to the Emperor, (b) embalmed, (c) tame, (d) sucking pigs, (e) sirens, (f) fabulous, (g) stray dogs, (h) included in the present classification, (i) frenzied, (j) innumerable, (k) drawn with a very fine camel-hair brush, (l) *etcetera*, (m) having just broken the water pitcher, (n) that from a long way look likes flies."[26] This is a long way from Linnaeus's taxonomy in which dolphins have to swim out of the box with fish into that of Mammalia, because they share lungs and teats with the latter and "just" water with the fish. Linnaeus's taxonomy is based on anatomy and physiology—not on the larger place or cultural practices in which the categorized entities are embedded and involved. Herman Melville's Ishmael in *Moby Dick* facetiously and perversely classifies whales (the taxonomic order to which dolphins belong) as fish, aware that there is controversy on the matter. Borges's Chinese taxonomy is grotesquely practice-, or, experience-based and thus defies "all the ordered surfaces and all the planes with which we are accustomed to tame the wild profusion of existing things." It can be faithful to the squirrels who live in the trees between earth and sky and defy as flying mammals all categories by defying gravity like the birds, crisscrossing horizontally and verticality, jumping between terrestrial and celestial animals.

It is no longer a question of reducing something to a sheer means versus an end, but a situation in which, as Bruno Latour would call it, there are only "mediators": "No entity is merely a means. There are always also ends. In other words, there are only mediators."[27] No origins, no ends/teloi, rather shifts, adjustments. Animals that defy our boundaries often become our obsessions: since we cannot domesticate (control) them, we either kill them or declare them sacred. Fish in Thoreau's Walden Pond symbolize the transcendental, as do birds. Where we humans are bound by vertical and horizontal boundaries of our bodily coordinates, fish and birds (and sometimes squirrels) crisscross space freely, diagonally.

Boundary Birds

Texas is the only state that has the honor of having its own Peterson bird guide: *Field Guide to the Birds of Texas and Adjacent States*. The rest of the nation is just anonymously divided into East and West: *Field Guide to All the Birds of Eastern and Central North America* and *Field Guide to Western Birds*.

This honor is well deserved, because Texas, lying on crucial bird migration routes, is a birding paradise. Texas is not just a border *state* with the Rio Grande separating the United States and Mexico, it is a boundary *object*, connecting arctic Canada with tropical Latin America.

Texas has 620 documented species of bird, considerably more than any other state. One of its most gorgeous birds is the scissor-tailed flycatcher, also called Texas bird-of-paradise. It was heavily poached because of the delicate color scheme of its feathers, which are soft salmon pink and light grey. But nowadays it is more seriously threatened by loss of habitat, the most significant cause of species extinction. Scissor-tails prefer grassland habitats with a few scattered trees, which function as their nesting sites and as places to perch while hunting for grasshoppers and other insects. Trees growing along fencerows are another favorite place for scissor-tails. When they are cleared or when the fencelines themselves are removed, the scissor-tails lose significant habitat. The biggest threat is encroaching subdivision of ranchland.

Another charismatic bird is the whooping crane. Its migration route runs from the delta of the Guadalupe and San Antonio rivers near Matagorda Bay in Texas to the deep Canadian Rockies. The population is migratory and winters in the Aransas National Wildlife Refuge on the Texas gulf coast. The Aransas range is protected, as is the whoopers' last breeding area in Canada's Wood Buffalo National Park. However, Aransas is next to an intercoastal waterway with heavy commercial ship traffic, much of it hauling petroleum products from nearby coastal refineries. Wildlife biologists are in the process of establishing new populations to ensure that the species will not be wiped out by an oil spill or other industrial disaster—or, for that matter, natural disaster—there. In January 2005, the whooping crane population was estimated at 342 birds.

But, like the scissor-tail, the issue of greatest concern for the cranes at Aransas is habitat. Their wintering grounds are federally owned, but the flow of freshwater into the Aransas National Wildlife Refuge is beyond the control of the federal U.S. Fish and Wildlife Service. However, the health and survival of the endangered whooping crane flock are directly related to freshwater inflow. It keeps the water drinkable for the cranes by maintaining the salinity levels low enough. Fresh water is also vital for the blue crabs, the primary food source for whooping cranes.

Scissor-tails and whooping cranes can serve Texans as boundary objects, facilitating the negotiation of land use and water rights, bridging business and environmental concerns. Nature tourism is a fast-growing segment of the tourism industry, delivering billions of dollars. It affords rural communities a low-impact way to diversify their local economies and agricultural incomes. Texas Parks and Wildlife began in 1993 the Great Texas Coastal Birding Trail, a 500-mile highway-train network that connected existing and new birding sites and linked birdwatchers (the users) with private landowners of ranches and community business such as restaurants, lodging establishments, and other providers of goods, all serving the birders. The whooping crane is one of its major attractions; it is the biggest bird native to North America. No wonder it wants to winter in Texas, because Texas is home to the biggest everything—Don't mess with Texas! Texas knows how to converse with coding tricksters; its state bird is the mocking bird.

At the Edge

"It is at the edge of the petal that love awaits," William Carlos Williams states, capturing aptly the potential of the edge through the transformative power of love.[28] I encountered this quote in a photo coffee table book, in which meaningful and wise quotes transported the beautiful pictures into deeper dimensions. When I googled the sentence to find its exact source, I did not see any reference to a particular poem or book of William Carlos Williams, but got lost in a plethora of commercial Web sites. Thinking I had found something unique, a rare gem, exclusive, exceptional, I had hit the ordinary, a widely over(?)-used quote, circulating expansively in a wide world of florist Web sites, advertising "creative designs, perfect sentiments"; greeting-cards sites; sites for bridal & wedding services, birthdays, and other fortunate occasions; garden-quotes Web sites; Valentine-Day–Love-Letter quote galleries: "If you find yourself at a loss for words to express your emotions, try the listing of quotations below. You may find just the right words." And Williams's love at the edge of the petal emerged as a favorite candidate for "right words." I saw thousands (millions?) of eyes staring at the computer, printing their own "original" card with the singular special quote about the edge, the petal, and love. Enticed by the special, I found the banal—a final oxymoron, a most appropriate place for an end, another edge. For, as I said at the beginning, the edge of boundaries lies in their transitional force. And love is a most powerful engine for transition or transformation. Because love takes the other seriously. Hence love is a never-ending practice, an ongoing invitation to change, to adjust to the other. Love is a boundary project, an edge effect, a transition zone. To love is to find the transitions in boundaries, not only of the charismatic rose but also of the greening of the grass. To love is to find the special in the everyday. Look: It is at the edge of the grass that love awaits.

Notes

I want to thank J. Baird Callicott for his masterful facilitating of textual transitions and Priscilla Ybarra Solis for her careful reading of various previous versions of this text.

1. Anne Michaels, *Fugitive Pieces* (Toronto: McClelland & Stewart, Inc., 1996), 140.

2. Ludwig Wittgenstein, *Philosophische Unersuchungen*. Frankfort am Main: Suhrkamp, 1971, #122, 82. "*Die übersichtliche Darstellung vermittelt das Verständnis, welches eben darin besteht, das wir die 'Zusammenhänge sehen.' Daher die Wichtigkeit des Findens und Erfindens von Zwischengliedern.*" I only partly follow G. E. M. Anscombe's translation. See Ludwig Wittgenstein, *Philosophical Investigations*, trans. G. E. M. Anscombe, 3rd ed. New York: Macmillan, 1968, #122, 49.

3. The dead zone is a large area (22,000 km^2 in 2002) in the Gulf of Mexico with an oxygen level too low for most aquatic species to survive. The oxygen depletion, called hypoxia, is mainly caused by nutrient enrichment from anthropogenic sources. This leads to eutrophication, that is, increased algal production and organic carbon in the ecosytem. See http://www. nos.noaa/gov/ and http://www.epa.gov/msbasin/taskforce/pdf/05 factsheetupdate.pdf.

4. Bruno Latour, "Circulating Reference," *Pandora's Hope, Essays on the Reality of Science Studies*. Cambridge: Harvard University Press, 1999, 24.

5. Susan Leigh Star and James R. Griesemeier, "Institutional Ecology, 'Translations' and Boundary Objects: Amateurs and Professionals in Berkeley's Museum of Vertebrate Zoology," 1907–39, *Social Studies of Science* 19; London: Sage, 1989, 387–420, 412.

6. I. J. Klaver, J. Keulaertz, B. Gremmen, and H. Van der Belt, "Born to be Wild; A Pluralistic Ethic Concerning Introduced Large Herbivores," *Environmental Ethics* 24 (2002): 3–21.

7. See I. J. Klaver and J. Donahue, "Whose Water Is It Anyway? Boundary Negotiations on the Edwards Aquifer in Texas," eds. Linda and Scott Whiteford, *Globalization, Water and Health: Resource Management in Terms of Scarcity*. Santa Fe: School of American Research Press, 2005, 107–127.

8. Mary L. Cadenasso, Steward T. A. Pickett, Katheleen C. Weathers, Susan S. Bell, Tracy L. Benning, Margaret M. Carreiro, and Todd E. Dawson, "An Interdisciplinary and Synthetic Approach to Ecological Boundaries," *Bioscience* 53:8 (August 2003): 719.

9. See Val Plumwood, *Feminism and the Mastery of Nature*. London & New York: Routledge, 1993, especially chapter 2, "Dualism: the logic of colonization," 41–69. See also Chris J. Cuomo, *Feminism and Ecological Communities*. London & New York: Routledge, 1998.

10. Aldo Leopold, *Game Management*. Madison: University of Wisconsin Press, 1933: reprint edition 1986, xxxi. (Emphasis in original.)

11. Arjun Appadurai, *Modernity at Large: Cultural Dimensions of Globalization*. Minneapolis: University of Minnesota Press, 1996, 7, 10–11.

12. Michel Foucault, "Kant on Enlightenment and Revolution," trans. Colin Gordon, eds. Mike Gane and Terry Johnson, *Foucault's New Domains*. London: Routledge, 1993, 12–18.

13. Jean-Luc Nancy, *The Sense of the World*, trans. Jeffrey Librett. Minneapolis: University of Minnesota Press, 1997, 5, 152–155. Judith Butler, "Universality in Culture," ed. Joshua Cohen, *For Love of Country, Debating the Limits of Patriotism*, Martha C. Nussbaum with Respondents. Boston: Beacon Press, 1996, 51.

14. Merleau-Ponty, *The Visible and the Invisible*, trans. Alphonso Lingis. Evanston: Northwestern University Press, 1968, 259.

15. See a further elaboration of this thought in my essay "Stone Worlds: Phenomenology on the Rocks," eds. Michael E. Zimmerman, J. Baird Callicott, Karen Warren, Irene Klaver, and John Clarke, *Environmental Philosophy: From Animal Rights to Social Ecology*, 4th ed. New York: Prentice-Hall, 2004, 347–360.

16. Rachel Carson, *The Edge of the Sea*. 1955; reprint, New York: Houghton Mifflin, 1998, 215.

17. Martin Heidegger, "Building, Dwelling, Thinking," *Poetry, Language, Thought*, trans. Albert Hofsttadter. New York: Harper Colophon, 1975, 154.

18. Donna J. Haraway, "Situated Knowledges: The Science Question in Feminism and the Privilege of Partial Perspective," *Simians, Cyborgs, and Women: The Reinvention of Nature*. New York: Routledge, 1991, 200–201.

19. Leslie Silko, *Ceremony*. New York: Penguin Books, 1986 (1977).

20. Ibid., 75.

21. Ibid., 74.

22. Ibid., 121.

23. Ibid., 126.

24. Ibid., 130.

25. Slavoj Žižek, *Tarrying with the Negative: Kant, Hegel, and the Critique of Ideology*. Durham: Duke University Press, 1993, 1.

26. Michel Foucault, *The Order of Things, An Archeology of the Human Sciences*. New York: Vintage Books, 1973, xv.

27. Bruno Latour, "To Modernise or to Ecologise? That is the Question," N. Castree and B. Willems-Braun, eds., *Remaking Reality: Nature at the Millenium*. London & New York: Routledge, 1998, 232.

28. William Carlos Williams, "The Rose (The rose is obsolete)," in *The Collected Poems of William Carlos Williams Volume I (1909–1939)*. New York: New Directions, 1986.

Chapter 8

Remapping Land Use: Remote Sensing, Institutional Approaches, and Landscape Boundaries

Firooza Pavri

An understanding of processes that drive forest expansion or degradation must necessarily be built on firm knowledge of institutional arrangements that govern forest use, a thorough understanding of particular people–forest interactions, and the systematic monitoring of particular sites of interest over long time periods. Across the rural regions of many developing countries, forests serve as important resource endowments, meeting local fuel wood, food, and timber requirements for many landless and marginal peasants (Jodha 1986; Pavri 1999). Additionally, the impact of forests on the earth's physical system (e.g., assisting in carbon sequestration, regulating regional climates) provides advantages that transcend political boundaries. Thus, the appropriate management and long-term sustainability of these natural resources are crucial, not merely in terms of providing ecological stability, but also for ensuring local livelihoods.

In recent decades, researchers and policy-makers have turned to detailed land use and land cover mapping exercises using satellite-imaging (or remote sensing—RS) technology to monitor change and understand the reasons for change.[1] Yet, linking the results of this technology with ground reality is far from seamless and poses a number of challenges. Moreover, the—at times—inappropriate linkage of image information to the ground obfuscates the complex trajectories of land use and land cover change observed across the developing world. In parts of forested India where rural livelihoods are closely intertwined with forest health, the appropriate use of satellite imagery to facilitate the implementation of resource management decisions could mean the difference between resource availability or scarcity. This chapter uses the case of India to offer a framework that links image information to the complex socio-ecological interaction patterns of forest use and extraction observed on the ground. In doing so, not only does it hope to contribute to recent "participatory RS" debates, but also move toward extracting the full potential of this technology.[2]

Research addressing the processes driving forest cover change adopts varied theoretical explanations. These range from neo-Malthusian accounts (Meadows et al. 1972; Myers 1991) to institutional frameworks that give due consideration to the regional political economy and avoid perfunctory population-size driven explanations for forest cover change (Bromley 1991). Specifically, institutional analyses illustrate the impact of institutional form on environmental outcomes through empirically rich studies documenting the actions of institutions and the normative arrangements prescribed by property-rights regimes in shaping the practices of dependent forest users (Leach et al. 1997; Robbins 1998). Notwithstanding theoretical disagreements, satellite image data are increasingly employed to monitor forest cover change by constructing spatially bounded land use/cover categories or classes based on image spectral reflectance values.[3] In the hands of a policy-maker or forester on the ground, these images provide vital information influencing land management decisions.

The opportunities presented by satellite technology to answer landscape-change questions offer great promise. With the exception of a few modeling efforts (Liverman et al. 1998; Moran et al. 1994), scant attempts have seriously considered merging these data in ways that might contribute to more rigorous and contextually appropriate understandings of land use/cover change. For instance, would linking satellite image land use/cover categories to institutional categories that dictate user practices on the ground provide details about management failures and successes and in turn lead to more ecologically sensitive ways of conceptualizing land use/cover (or management) boundaries? And how might this linkage be executed? Simply put, can remote sensing, in fact, contribute to a fuller understanding of landscape change and lead to better management strategies, rather than primarily serve as a monitoring tool? Studies have attempted to address these concerns by exploring how image analysis might fine tune land use/cover categories such that we move away from generalized depictions of change (e.g., forest to agriculture, or rural to urban, etc.) to recognize subtle within-category variations (Lambin 1999). Still others have attempted to link image pixels to actual users on the ground through Global Positioning Systems (GPS)/Geographic Information Systems (GIS) technology (Liverman et al. 1998).[4] These attempts, and recent work by institutional ecologists (Rangan 2000; Robbins and Maddock 2000), open the door to an institutionally sensitive land use/cover remote sensing classification framework.

Imaging Technology and Landscape Boundaries

Remote sensing technology has experienced remarkable advances since early balloon-hoisted cameras captured images of the landscape below. Since the early 1970s, space-based satellite sensors have efficiently transmitted greater amounts of information from 900 kilometers in space than that obtainable through aircraft mounted sensors closer to the earth's surface. By providing views of the earth's surface from wavelengths beyond the visible spectrum, these new data have also been instrumental in speeding up work on international environmental policy frameworks.

Land use/cover have long since been recognized as essential variables, key to understanding an ecosystem's dynamic functioning, and crucial to regulating flows that impact the biosphere (Skole et al. 1993; Turner et al. 1995). Advances in remote sensing enable systematic global land use/cover mapping exercises at temporal frequencies ranging from data collection every few days to every few weeks. The characteristics of land use/cover information obtained through this technology are determined by the unique combination of temporal frequency of image acquisition and the spatial and spectral resolution of individual pixel data. The selection of each depends on the study's purpose and the coverage needed to aid landscape pattern and process analysis (Cihlar 2000). For the most part, systematic and frequent temporal acquisition of images is instrumental when monitoring forest cover change.

Satellite data that provide high spatial resolution images allow for more detailed spatial pattern analysis, and supply greater amounts of information on surface conditions, including shape, size, edges, and other metrics for a particular land feature under scrutiny. For geographers, analyzing spatial patterns could also provide cues as to the causes of change. For instance, the extent of forest fragmentation might reveal the type of clearing practices employed, that is, small-scale logging or more extensive clearing, while a comparison of several images covering the same area but on separate dates would reveal ensuing land use practices. Reading landscape change based on the spatial resolution of an image results in a qualitative interpretation and depends, in large part, on the abilities of the analyst, his or her familiarity with the setting, and the availability of ancillary data such as climate variables to enhance interpretation. Studies, however, rarely rely on just spatial resolution for interpretation. Rather, they incorporate this information with more quantifiable spectral data analysis.

Image spectral data provide detailed landscape information by capturing the energy (measured electronically for different wavelengths or bands by the satellite sensor) reflected from a land surface feature (Lillesand and Kiefer 2000). Each satellite sensor is uniquely designed to record reflected energy from specific portions of the electromagnetic spectrum and, in turn, reveal the reflectivity characteristics or unique spectral signatures of land features.[5] Computer-assisted statistical techniques ultimately allow for a fuller examination of spectral patterns (Lillesand and Kiefer 2000). Researchers interested in land use/cover change have employed image spectral data to classify similar image pixels into appropriate land use/cover classes. These classification routines form the basis for all land use/land cover change studies, and the information they provide are vital to management and policy decisions (Robbins and Maddock 2000).

Unique combinations of spatial, spectral, and temporal components of image data arm analysts with the overall monitoring framework and eventually provide detailed representations of landscape change patterns. Remote sensing thus offers a systematic record of land transformation, purporting to rely on objective and reproducible results. Unlike customary landscape data collection procedures, images are relatively free of political baggage, although they necessarily reflect some subjectivity inherent to most data classification procedures, whether in the actual naming of classes or the type of procedure used (Robbins and Maddock 2000). Moreover, as Robbins and Maddock suggest, land cover categories realized through image analysis can be problematic when imposed unquestioningly on already existing local land cover classes, obliterating locally significant measures of landscape features. Recognizing the divergent ways in which different groups of people identify and use land cover categories can be vital to offering appropriate management strategies. Participatory frameworks incorporate alternative forms of spatial knowledge representation, and organize conflicting views, both "expert" and "local," such that these inform our understanding of landscape change and lead to more productive land management policies (Harris and Weiner 1998; Omotayo and Musa 1999).

Exercises in participatory land cover mapping question "expert" interpretations of landscape change and are a step in the right direction. They do not, however, help us extract the full potential of satellite imagery. Besides pointing out differences in the way groups perceive the landscape and providing a voice to those on the periphery, these counter mapping exercises have yet to offer methods that will appropriately link data generated by images to those obtained through ground survey exercises or participatory remote sensing. To this end, studies from the Brazilian Amazon engaged in mapping land use trajectories after deforestation have accomplished much in terms of linking patterns observed through images to locally/regionally contingent factors influencing ensuing land use decisions (Liverman et al. 1998; Moran et al. 1994). Even so, this literature has yet to consider how interpretations of satellite land cover mapping results might be aided by an understanding of the role of institutions in mediating land use practices. Rather, land use mapping exercises need to directly engage in an understanding of how institutional boundaries are arrived at, how they impact land use practices, and what effect they have on the landscape.

Institutional Boundaries and the Indian State

Institutional frameworks have demonstrated that most contemporary society–environment interactions are mediated by institutions that legitimize systems of rights and rules governing how, and by whom, available resources are used (Bromley 1991; Feeny et al. 1998; North 1990; Ostrom 1990). Property-rights institutions (whether private, common, or state-managed), give rise to landscape boundaries by exercising control over, and governing access to, resources. Through resource rights and rules, these regimes organize the behavior patterns of users to produce resource use practices that invariably impact the quality and quantity of the resource itself (Hanna et al. 1996).

The contemporary Indian Forest Department was set up by the colonial government in 1864 and charged with managing the country's extensive forest reserves. In an attempt to control the country's vast and heretofore unfamiliar forest reserves, the colonial government initiated scientific forest management based on the work of German foresters of the time (Grove 1995). Scientific management claimed to promote more ecologically sound practices. These included the meticulous categorization of forests based on their production capabilities and the discouragement of local forest use practices (e.g., shifting cultivation) or forest institutions (e.g., common pool resource institutions) deemed ecologically risky (Gadgil 1991). Early state intervention in forestry was influenced by several motives and reflected the growing political and fiscal imperatives of the day. Aside from economically driven motivations, early Forest Department memoranda also rationalized securing forests under state control evoking the state's superior ability to achieve ecological and economic objectives while meeting the fuel wood needs of rural populations (Bombay Revenue Department 1906; Rangan 2000). This early period of land consolidation produced a categorization system of forest access regimes, which included Reserved, Protected, Unclassed, and Village forests. Each category was bound by different management systems, rights and rules of access, and local or regional authorities that controlled forest use. Forest management in independent India was similarly adjusted to meet contemporaneous demands and while original forest categories were largely untouched, cosmetic changes were made to access conditions and rights of use.

Today, as before, control over access is generally vested with a local or state authority and while there is little flexibility in prescribed access rights and rules of use on paper, the reality of forest management is a process of negotiation between the authority and local users that plays out at particular sites across the country. Regular adjustments to formally prescribed rules and sanctions for violations are made to fit individual contexts. Discretion is exercised on the basis of a complex equation that includes informal familial ties, politically motivated favors, and deference to important local families (Pavri and Deshmukh 2003). In some cases, inadequate regulation and enforcement owing to a shortage of personnel, or the breaking of formal rules, or corruption among forest managers leads to eclectic patterns of both local and commercial use (Pavri 1999). Thus, variations in the degree of control an authority can exert and the ability of local users to negotiate prescribed access conditions through legal or extralegal means have direct implications for observed land use practices, in turn resulting in a diversity of land use/cover types.

How might remote sensing technology help us study these patterns of use? Clearly, land use/cover change studies can be informed by examining eclectic forest landscape patterns resulting from diverse institutional forms. Rather than concentrating on categorizing land use/cover into discrete classes that observe internal homogeneity as has been the case, analysts might try to uncover the diverse use patterns through image analysis and link these to institutional practice and local conditions of use. Satellite imaging technology has provided social scientists with an incredible wealth of land surface information. Even so, in their attempt to unravel complex landscape patterns, conventional methods disregard the equally important social context contributing to particular land use/cover use patterns. In areas where multiple land users and land uses exists, the appropriate interpretation of landscape change hinges on not just a firm understanding of the physics of land surface reflectivity and satellite data capture. Rather it centers on understanding local contexts, institutional form, and overall society–forest interaction patterns and their ensuing impacts.

Thus far, studies in geography have kept these two sources of information separate. On the one hand, satellites provide quantifiable land use change statistics, while data from the ground allow for an assessment of contributory factors. In a move to bridge the two and provide integrated modeling efforts, spatially explicit models of human behavior have attempted to represent landscape change based on linking pixels to people (Liverman et al. 1998). Using high-precision digital mapping tools, image information is combined with that obtained from the field. Few attempts have linked these with the intention of understanding underlying institutional regimes that facilitate land use patterns. Even so, these studies have expended much effort on data linkage techniques and thus provide a step in the right direction.

Using high-precision digital mapping technology and field information, institutional boundaries might be etched onto an image once appropriate image processing techniques have been conducted. Extensions to conventional land use/cover categorization methods, which include fuzzy clustering and textural analysis, provide especially useful ways of conceptualizing land use categories.[6] Such dual classification methods (both spectral and institutional) would serve a variety of functions and make for more effective ways in which the appropriateness of particular land use practices might be assessed. First, through fuzzy classifiers, one could move away from unitary categories such as "forest" and "grassland" to ones that incorporate subtle variations that typify eclectic landscapes resulting from multiple use practices.[7] Unlike conventional classifiers, these would prevent valuable image information from being lost. For instance, it would be more appropriate to categorize a sparsely grass covered region as precisely that, rather than lumping it together with either bare soil or grass. Subtleties in categorization could have special impacts for the types of management practices employed. To use the same example, a sparsely grass-covered region would be managed very differently if it were considered within the soil category as opposed to grassland or vice versa. Fuzzy classification allows for more effective representations of complex land cover conditions (Wang 1990).

Second, overlaying institutional categories on images would allow for more transparent analyses of how different institutional governance strategies, and the techniques they employ to control use, impact the landscape. Inevitably, this kind of analysis would direct attention to management failures and successes—clearly of value to an analyst monitoring change. While it might not reveal all the causal processes behind change, this type of institutional overlay analysis would document areas of concern, and be more sensitive to issues of property, control, and authority to interested institutional researchers.

Third, and using the capabilities of high precision mapping, one might then start merging or linking the information arrived at through fuzzy classifiers with that obtained through field surveys and institutional boundaries. The resultant landscape would be an eclectic mix of spectral and institutional categories, and could further be enhanced with information gathered through interviews with actual resource users themselves. These data might include information on land use practices, or other subjective reasons why local users chose particular extraction practices. In so conducting image analysis, we arrive at more appropriate ways in which to extract the full potential of satellite data while simultaneously recognizing and linking these to social processes that impact land use change. Finally, this type of a linked satellite image-institutional analysis would open the door to more locally aware and institutionally appropriate management alternatives.

Notes

1. Land cover has been defined as "the composition and characteristics of land surface elements" and is generally determined by a combination of climatic factors and soil conditions (Cihlar 2000). Land use, on the other hand, refers to how society uses a particular land cover type.

2. Participatory or community-integrated Geographic Information Systems (GIS)/RS are relatively recent thrusts that purport to incorporate local knowledge and expertise in GIS-related or remotely sensed land use/cover analyses. Practitioners point to the potential of a democratized version of these technologies in empowering communities and representing the worldviews of those typically on the periphery.

3. Satellite-mounted sensors provide remotely sensed images of the earth's surface by recording energy reflected off land features at different wavelengths of the electromagnetic spectrum (EMS). Thus, in addition to recording information from the visible spectrum, satellites sensors also record information in the near, mid, and far-infrared (IR), and thermal wavelengths (bands) of the EMS. These data result in "typical" spectral responses patterns, or signatures, for different biotic and abiotic land features (Lillesand and Kiefer 2000).

4. A pixel, or picture element, constitutes the smallest observable spatial unit on a remotely sensed image. The spatial resolution of a satellite sensor determines the extent of land represented by a particular pixel for that sensor. For example, the popular Landsat Thematic Mapper sensor has a pixel spatial resolution of 30 m sq, where each pixel represents data collected from a 30 x 30 m land surface area.

5. For instance, Landsat Thematic Mapper captures data in seven wavelengths (or bands), including the visible portions (blue, green, red), the near, mid, and far-infrared, and thermal sections of the EMS. The spectral values (from each pixel in each band) that an image analyst works with on the computer are merely positive integers (also called Digital Numbers, *DN*) usually ranging from 0 to 255, obtained by an electrical signal to digital conversion process (Lillesand and Kiefer 2000). Simple interpretation would follow thusly: a pixel with a value of 45 in the blue band records a land feature that absorbs energy within the blue portion of the EMS, while a pixel value of 197 in the same band would record a land feature that reflects blue light or energy.

6. Fuzzy classifiers are especially useful for coarse spatial resolution data and take into consideration mixed pixel issues—i.e., where one pixel might contain two or more land cover classes because of the large land area that the pixel represents (Wang 1990). Textual analysis, on the other hand, includes a visual interpretation of image texture into classification procedures. Image texture generally relies on identifying the

roughness or smoothness of image features. For instance, open forest canopy cover might be of a rougher texture than a pre-harvest wheat field.

7. While unitary categories might work in some cases, namely, the Amazon basin, for other parts of the developing world, where land use categories mix and blur clearly defined boundaries, fuzzy classifiers would be able to capture landscape features more effectively.

References

Bombay Revenue Department, 1906. *Forest Proceedings. Note from the Government of India to the Finance Dept.*, no. 389, 1 November 1906.

Bromley, D. 1991. *Environment and Economy: Property Rights and Public Policy*. Cambridge: Basil Blackwell.

Cihlar, J. 2000. "Land Cover Mapping of Large Areas from Satellites: Status and Research Priorities." *International Journal of Remote Sensing* V21(6 and 7): 1093–1114.

Feeny, D., F. Berkes, B. J. McCay, and J. Acheson, 1998. "Tragedy of the Commons: Twenty-two Years Later." *Human Ecology* 18: 1–19.

Gadgil, M. 1991. "Deforestation: Problems and Prospects." In A. S. Rawat, ed. *History of Forestry in India*. New Delhi: Indus Publishing Co.

Grove, R. 1995. *Green Imperialism*. Cambridge: Cambridge University Press.

Hanna, S., C. Folke, and K-G Maler, 1996. "Property Rights and the National Environment." In S. Hanna, C. Folke, and K-G. Maler, A. Jansson, (eds.), *Rights to Nature: Ecological, Economic, Cultural, and Political Principles of Institutions for the Environment*. Washington, DC: Island Press, 1–10.

Harris, T., and D. Weiner. 1998. "Empowerment, Marginalization and 'Community-Integrated' GIS." *Cartography and Geographic Information Systems* 25(2): 67–76.

Jodha, N. 1986. "Common Property Resources and Rural Poor in Dry Regions of India." *Economic and Political Weekly* 21(27): 1169–1181.

Lambin, E. 1999. "Monitoring Forest Degradation in Tropical Regions by Remote Sensing: Some Methodological Issues." *Global Ecology and Biogeography* 8: 191–198.

Leach, M., R. Mearns, and I. Scoones, 1997. "Environmental Entitlements: A Framework for Understanding the Institutional Dynamics of Environmental Change." IDS Discussion Paper #359, IDS, Brighton.

Lillesand, T. M., and R. W. Kiefer. 2000. *Remote Sensing and Image Interpretation*. New York: John Wiley and Sons.

Liverman, D., E. F. Moran, R. Rindfuss, and P. Stern (eds.). 1998. *People and Pixels: Linking Remote Sensing and Social Science*. Washington, DC: National Academy Press.

Meadows, D. H., D. L. Meadows, J. Randers, and W. W. Behrens III. 1972. *The Limits to Growth*. New York: Universe Books.

Moran, E. F., E. Brondizio, P. Mausel, and Y. Wu. 1994. "Integrating Amazonian Vegetation, Land-Use and Satellite Data." *Bioscience* V44(5): 329–338.

Myers, N. 1991. *Population, Resources and the Environment*. London (UNFPA): Banson Production.

North, D. 1990. *Institutions, Institutional Change and Economic Performance*. Cambridge: Cambridge University Press.

Omotayo, A., and M. Musa. 1999. "The Role of Indigenous Land Classification and Management Practices in Sustaining Land Use System in the Semi-Arid Zone of Nigeria." *Journal of Sustainable Agriculture* 14(1): 49–58.

Ostrom, E. 1990. *Governing the Commons: The Evolution of Institutions for Collective Action*. Cambridge: Cambridge University Press.

Pavri, F. 1999. "Tragedies in State Commons: Macro Forest Policies, Local Influences and Deforestation in the Western Ghats of Raigad, India." Ph.D. diss., Ohio State University, Columbus, Ohio.

Pavri, F., and S. Deshmukh. 2003. "Institutional Efficacy in Resource Management: Temporally Congruent Embeddedness for Forest Systems of Western India." *Geoforum* V34(1): 71–84.

Rangan, H. 2000. *Of Myths and Movements: Rewriting Chipko into Himalayan History*. London: Verso.

Robbins, P. 1998. "Authority and Environment: Institutional Landscapes in Rajasthan, India." *Annals of the Association of American Geographers* 88(3), 410–435.

Robbins, P., and T. Maddock. 2000. "Interrogating Land Cover Categories: Metaphor and Method in Remote Sensing." *Cartography and Geographic Information Science* 27(4), 295–309.

Skole, D., B. Moore III, and W. H. Chomentowski. 1993. "Global Geographic Information Systems and Databases for Vegetation Change

Studies." In A. M. Solomon and H. H. Shugart. eds., *Vegetation Dynamics and Global Change*. New York: Chapman and Hall, 168–189.

Turner, B. L., D. Skole, S. Sanderson, G. Fischer, L. Fresco, and R. Leemans. 1995. *Land Use and Land Cover Change: Science/Research Plan*. Stockholm & Geneva: International Council of Scientific Unions.

Wang, F. 1990. "Fuzzy Supervised Classification of Remote Sensing Images." *IEEE Transactions on Geoscience and Remote Sensing*, V28(2): 194–201.

Chapter 9

Boundaries, Communities, and Politics

Anna L. Peterson

This chapter looks at how two different social groups define, experience, and often blur a series of boundaries: between individuals and society, between human and natural communities, between the local and the global, and finally between mundane and sacred realities. The two groups are both religiously grounded and agrarian, one in El Salvador and the other in the midwestern United States. Both are relatively homogenous in terms of religion, culture, and ethnicity. They live in a face-to-face world where community members know each other. These communities also share certain values supported by a religious underpinning. Especially important is their mutual rejection of excessive individualism and their understanding of human nature as essentially social. Comparing these two communities, both struggling to embody socially and ecologically sustainable values, highlights important common structures and values that might shape other models of socially just and environmentally sustainable societies. They also point to the ways in which religion complicates and also enriches discussions of natural and social boundaries.

Catholic Peasants in El Salvador

The changes in Catholic pastoral work and theology following the Second Vatican Council (1962–1965) had a sweeping, though varied, impact throughout Latin America. Reforms aimed to democratize church structures, to educate and empower lay leaders, and to commit the church to social justice. The primary vehicle for these changes were *comunidades de base* (CEBs):

small groups that met weekly to discuss biblical readings in light of their own experiences. In many parts of Latin America, including El Salvador, CEBs and the laypeople who participated in them became linked to opposition social movements. These movements expressed their values and aspirations in religious language: to create a "new man," to begin building a "Christian society," and to make their communities embody the harmony and justice that characterize the reign of God.

In El Salvador, progressive Catholic ideas strongly shaped peasant organizations beginning in the late 1960s. Groups such as FECCAS (the Federation of Catholic Peasants) became increasingly powerful and militant throughout the 1970s, and faced intense repression as the military government cracked down on opposition activism late in that decade. By 1981, El Salvador was in the midst of a full-scale civil war between the government and the guerrilla combatants of the Frente Farabundo Marti para la Liberacion Nacional (FMLN). Tens of thousands of peasants fled the war zones for church-run refugee camps in San Salvador or, as the war dragged on, in camps set up by the United Nations over the border in Honduras. In the camps, refugees organized literacy classes, taught themselves skills like carpentry and sewing, and formed groups to press for their rights, for example, to be free from harassment in the camps by the Salvadoran and Honduran armies. By the mid-1980s, many refugees began planning to return to El Salvador, even though the war continued. The first, small repopulation in 1985 was followed by a series of larger, well-organized moves from camps in Honduras to abandoned villages in war zones, especially northern provinces such as Chalatenango, Cabañas, and Morazán.

I focus here on Chalatenango, a center of both peasant organizing and government repression. I first visited repopulated villages there in October 1988, a year after the first repopulation, to Guarjila, and a few months after the second one, to nearby San Antonio los Ranchos. At that time, people in Guarjila lived in mud and stick huts with tin roofs and had no running water, electricity, or other infrastructure. Their collective energies for the first year had gone to planting and harvesting the beans and corn they needed to survive and, as their next priority, to constructing a school—the reborn village's first permanent building. Conditions in one-year old Guarjila were luxurious in comparison to those in San Antonio Los Ranchos, whose inhabitants spent all day clearing weeds from fields so they could plant basic crops and slept under plastic tarpaulins tied to sticks or the skeletons of bombed-out buildings. It was hard to imagine, at that point, how these people—mostly women and children, joined by some old folks and young men disabled by land mines or combat—were going to survive, much less fulfill their high hopes of constructing communities that would serve as models for a "new El Salvador."

They did survive, and they were joined by thousands of returned refugees who resettled other villages throughout the late 1980s. In 1992, the war ended as a result of negotiations brokered by the United Nations. After over a decade of peace, the repopulations continue to struggle on many fronts, but their achievements are astonishing. Every village is governed by a democratically elected council, which answers to the community as a whole. Economically, the repopulations have moved away from the nearly complete collectivization of ownership, production, and distribution that they adopted during their formative years, but most important resources, including cottage industries, are still commonly owned and their profits are commonly distributed. Their workers are employees of the community, as are health care providers, who run clinics that serve surrounding hamlets as well as the resettlements, and teachers.

Social changes are accompanied by evolving environmental practices. Most repopulations have made collective decisions to set aside land that is not to be farmed and have reforested selected areas. A number of individual farmers have begun terracing and crop rotation, lessened their use of chemical pesticides and fertilizers, and abandoned post-harvest burns. A few are practicing organic farming, particularly in the Lower Lempa region, where the wartime abandonment of farmlands and orchards facilitated the process of organic certification. With support from national and international NGOs, many communities are also pursuing alternative energy sources. In Guarjila, for example, residents dry locally grown fruit with a solar dryer; solar lighting is also used in the building in which the fruit is processed and packaged.

Not all these efforts are successful. Some have started and failed, others have undergone major transformations in order to survive economically, and still others manage only with substantial foreign aid, although this has fallen since the war's end. Many residents now work outside the repopulations, and the scarcity of jobs and low incomes remain chronic concerns, especially when they drive young people away from the villages. Still, most remain, held by family ties and an ongoing commitment to the collective vision of a new society. They reflect a unique vision of reinhabitation, an effort to live sustainably in poverty, on impoverished land, in the aftermath of war and political violence.

The Old Order Amish

The Old Order Amish, like all Anabaptists, are distinguished by their vision of how Christian beliefs ought to be lived out. From its origins, the Anabaptist movement called for a community of true believers apart from the established church, which, Anabaptists believed, had compromised with secular power and become corrupt and ineffectual. Anabaptists insist that the church must embody, in collective fashion, the values and characteristics of the early church and even of the reign of God—not just hope for their future realization, but live them out communally here and now. This Christian community will always be a minority, Anabaptists acknowledge, but this makes their obligation no less binding. The church is literally a "city on a hill," separate from the corruption and violence of the world that lives by the "sword," testifying to and embodying the alternative values of the future reign of God.

Old Order Anabaptist groups such as the Amish have sharpened Anabaptists' insistence on spiritual and physical separation between the church and the world. The Amish emerged out of the Anabaptist movement of the sixteenth and seventeenth century, and the first Amish migrated to North America in the early eighteenth century. Today, most of the 180,000 Old Order Amish are in Pennsylvania, Ohio, and Indiana, while smaller groups reside in other states, Canada, and Latin America. Despite differences among these settlements, all are driven by the struggle to separate from the world and create communities in accord with their religious principles. Distinctive features of Amish settlements include horse and buggy transportation; use of horses and mules in fieldwork; plain dress (in many variations); beard and shaven upper lip for (married) men; prayer cap for women; Pennsylvania German dialect; worship in homes; eighth grade private schooling; rejection of electricity from public utility lines; and prohibition of ownership of televisions and computers.[1] These markers are outlined and upheld in each community by its *Ordnung* (Order), a usually unwritten set of rules, expectations, and common values. Amish communities today remain largely agricultural, although the percentage that earns a living by farming has fallen significantly since World War II, due to shortages of affordable farmland and high Amish population growth. Still, most Amish families engage in occupations that keep them close to their families, such as home-based cottage industries and shops.

The church district is the basic Amish social unit, shaping virtually all aspects of members' lives. Church leaders make decisions about applying the rules, developing new ones, allowing for exceptions, and disciplining violators. However, their power is not absolute. All leadership posts and responsibilities are decided by community members, and there are no regional or national hierarchies. The *Ordnung* is decided collectively, and preachers and bishops are called by the community and receive no payment. Since face-to-face contact is frequent, it is not hard to hold leaders accountable for their decisions. Proximity is made necessary by the rejection of automobile ownership, one of the many ways in which the rejection of cars reinforces Amish values. Reliance on horse-drawn equipment keeps farms small and pushes different generations to live near each other whenever possible. Along with the church district, the family is the most important Amish social unit.

Generations live and work together, and children are educated largely at home, attending school only through eighth grade in Amish-run schools populated and staffed largely by relatives and neighbors.

Shared religious commitments, strong extended families, and locally based economies all help make Amish communities socially and religiously "sustainable"—able to last over time without losing core values. They are also ecologically sustainable, due in no small part to the prohibition on motor vehicles instituted early in the twentieth century. Amish farmers, while not all organic, rely much less heavily on chemical inputs and fossil fuels than conventional farmers. They have a long-standing commitment to crop rotation and diversification and the use of animal manure as a primary fertilizer, and farms stay small because there are limits to what a family can work with horse and human power.

Today the Amish face many challenges, most notably the difficulty of acquiring and keeping good farmland. Rising land prices have encouraged some farmers to sell land in their traditional strongholds and begin farming elsewhere, prompting the formation of new church districts in such states as North Carolina. Many Amish communities have experienced rapid growth in nonfarming occupations, including some that are closely related to agriculture and others that involve work in factories or shops.[2] Despite these challenges, predictions of the Old Orders' imminent demise have been spectacularly wrong, as the communities have grown in recent decades as a result of high birth and retention rates.

Although they face ongoing challenges, the Amish may be the best North American proof that it is possible to live collectively by values that diverge in almost every respect from those of the mainstream. They also offer some lessons about how this is possible. It requires substantial and increasing participation in national and regional economies, a certain degree of political organization and allies, and cooperation from the state, as well as a deeply rooted set of values to which members are uncompromisingly committed.

Common Frameworks, Boundaries, and Political Change

Bases of Community

These two groups, North American Amish and Salvadoran Catholic, diverge in numerous ways. However, they also share a number of common bases, both structural and ideological. Some are obvious: both are rural, agrarian communities, largely made up of small farmers, although this is changing as new forms of livelihood become necessary in both kinds of communities. Further, the communities are fairly homogeneous in terms of religion, economic status, and ethnicity. And these are small communities, whose members know each other and interact face-to-face all the time. In addition to these characteristics, we can identify commonalities in the values and worldviews that hold these communities together.

First, both the Amish and the Salvadoran Catholic repopulators understand human nature as essentially social. In both principle and practice, they reject the hyperindividualism of mainstream U.S. culture and instead understand themselves primarily in relation to larger collectives. Persons are not autonomous, self-sufficient individuals but rather social beings who can develop and thrive only in the course of mutual interactions and dependencies with others. This understanding has concrete consequences: social and natural resources are organized and distributed to benefit the group. This makes possible survival of individual members and of the collective. It also enables them to define "success" according to the well-being of the whole, not the fate of the privileged.

Second, these convictions have religious underpinnings that are vitally important, and in fact religion may be the most important base of each community. While this is more obvious for the Amish, it is equally true of the repopulations. Their understanding of themselves as individuals and community members, their vision of what a community can and should do, and their persistence in pursuing this vision all stem from a profoundly religious commitment, shaped by uniquely Latin American and Salvadoran experiences. Religion relates worldly experiences, aims, and standards to ultimate values and ends; it makes people part of something larger, better, and longer-lasting than themselves. It also serves practical purposes: it shapes everyday patterns of behavior, forms of interaction among members and between members and outsiders, economic and political organization, and priorities for social agendas. Both Amish and progressive Catholics faith is marked by an understanding of the reign of God as a model for contemporary social life, not merely a future aspiration.

One more base of community is central in both cases: they are rooted in particular places and defined by their knowledge of and relations to these places. Finding a place to settle down has cost both former Salvadoran refugees and the Amish dearly, and members of both groups retain an awareness that exile could happen again. Their attachment to and knowledge of place is deep-rooted, but also self-consciously vulnerable. Their ethics are thus both cosmopolitan and provincial, as Holmes Rolston puts it,[3] grounded in local commitment and experience at the same time they are oriented toward larger spatial and temporal contexts.

Nourishing Common Frameworks and Boundaries

What nourishes community in both cases is the ongoing effort to live out religious values. One of these values is what progressive Catholics call the preferential option for the poor and Anabaptists call mutual aid. The underlying principle in both cases is that no community member should be left behind. Thus, care for the most vulnerable community members, such as small children, the elderly, disabled, orphans, and widows, is a high priority in community institutions and practices. Catholic social thinking has made this principle explicit in reference to national and international economics, where the common good and the well-being of the least well-off should have priority over the pursuit of economic growth or individual profit.

These communities do not meet all their members' needs, eliminate every injustice, or transcend the limitations of their economic and political conditions. Still, they are far ahead of mainstream U.S. or Latin American societies in their attention to their members' needs. For example, the returnees in El Salvador made the construction of schools and health clinics one of their top priorities upon arriving back in El Salvador. Only planting crops necessary for survival took precedence; that done, they built schools even before starting permanent houses. Other high priorities have been the construction of health clinics and the training of health care providers. Today several repopulations have centers for elder care as well. Vulnerable community members, such as single parents, orphans, and the elderly, also receive collective support in the form of volunteer labor, food, and other supplies. All these projects are initiated and sustained at significant cost by community members who themselves have very scarce resources. Although the Amish do not live in the profound poverty and scarcity that shape the lives of the Salvadoran repopulators, with much less "spare" time and money than middle-class urban and suburbanites, they also have made concern for the least well-off a priority. Anabaptist practices of neighborly mutual aid and lifelong care for the elderly and disabled embody these traditional values.

Another value that nourishes these communities is popular participation. Although the Catholic Church remains hierarchical in its authority structure and decision-making procedures, the progressive Catholic vision aims for substantive lay participation in decision making at local and larger levels of the church and, especially, for democratization of economic and political structures. These goals are evident in the egalitarian political structure of the repopulated communities. Repopulators see these structures and practices as part of their larger struggle for the democratization of Salvadoran society as a whole. The Amish, rooted in the Anabaptist tradition, lack the hierarchical ecclesial structure of the Catholic Church. They insist on mutual decision-making and discipline and define the church as a community in which all members have the right and the obligation to participate in making and carrying out collective decisions, and all are accountable for each other and for the group as a whole.

Both communities also aim for greater environmental sustainability. In recent years, the Catholic Church has been rethinking attitudes toward the natural world, as evidenced in statements from the pope, regional and national episcopacies, and individual theologians. Their reflections are based on the long-standing Catholic insistence that creation is for the common good and should be used to uphold the dignity of each person rather than to pursue individual profit, and that people must be good stewards and not exploiters of the natural world. This echoes Anabaptist attitudes, according to which both human and nonhuman nature is flawed, interdependent, and intrinsically valuable, all ultimately dependent on and subordinate to God's will. These perceptions have shaped the Anabaptist understanding of agriculture as not only a way to make a living, but also a contribution to the common good, the local community, and the land itself. This vision, in turn, rests upon a historical and theological conception of farming as a religious calling.

Both the Amish and the repopulators are primarily agrarian, and much of their environmental concern is related to agricultural methods and livelihood. In particular, the Amish have probably the most consistent record of sustainable agriculture in North America. They continue to farm without tractors and with a high percentage of organic inputs, although many do use commercial fertilizers. This is part of an overall lifestyle that does not depend on fossil fuels, does not contribute to urban sprawl, and leaves a relatively small ecological footprint. Long residence in the same place also gives many Anabaptist farmers knowledge of and appreciation for local species. This ecologically healthy lifestyle is tied to a deliberate rejection of consumerism, which the Amish fear would displace values such as community well-being and faithful discipleship. In the repopulations, low consumption is tied to poverty. Few residents have cars or telephones because few can afford them, and the infrastructure making them possible is largely absent. Still, many deliberately reject mainstrain U.S. values, especially the pursuit of comfort and profit at the cost of survival for the poor. These are seen as values of the Salvadoran elites, whose unwillingness to give up luxuries for the collective good creates ongoing social inequities.

These communities' experiences and valuation of nature are inseparable from their experience of making a living and also from their concerns about human health, livelihood, and social justice. This might lead environmental activists and philosophers in the United States and Europe, especially those sympathetic to deep ecological perspectives, to define these groups as "anthropocentric." Two cautions, however, must be raised in response. First, the anthropocentric–nonanthropocentric dichotomy is far from absolute. Many Salvadorans value species and landscapes that are not useful for humans. Community members speak animatedly about formerly rare or absent bird species they now observe regularly because of their reforestation efforts, and old people take pleasure in seeing species they remember from their youth and sharing their knowledge with younger people, who interpret this in relation to concepts such as biodiversity and endangered species. At the same time, they often feel they have to override these "nonanthroopocentric" values in order to survive, as in their continued use of some pesticides that are dangerous to seed-eating songbirds. Their situations and attempts to balance various goods, human and natural, thus illuminate struggles common to environmental justice movements in the United States, Latin America, and elsewhere.

Second, discussions of anthropocentrism and nonanthropocentrism (or eco or biocentrism) leave out the theocentric perspective that is central to both the Amish and Salvadoran Catholics discussed here. They view the natural world as God's creation, and the ongoing life of that creation and all its inhabitants is centered not around humans or any other created element but rather around the creator God. They struggle for environmental preservation, sustainability, social justice, and economic security in light of their overarching religious commitments, which set the context for their understanding of human and ecological values. Their experiences remain relevant, I believe, for philosophical discussions of natural value, but we should not make the mistake of trying to cast these communities into secular philosophical categories.

Relations between Common Boundaries and the Development of Common Political Goals

Religion plays a central role in linking local communities to larger traditions of thought, institutions, and networks. Progressive Catholics in Latin America have developed explicit political goals, grounded on their interpretations of the Bible, Catholic social teaching, and "the signs of the times." Progressive Catholics feel called to work actively in the world to co-create the reign of God in history, in which social life is organized to preserve the dignity of all persons and the common good, for which all creation is intended. Rural organizing in El Salvador since the 1960s has linked religious visions of human dignity and brotherhood to political goals of democratization and economic justice. This rich blending of Christian and radical politics was the dominant worldview of the Salvadorans who fled and later repopulated conflicted areas. They understand themselves as the model for a new society in El Salvador, as reflected in the motto of Ciudad Segundo Montes, a repopulation in Morazán province: "a hope that is born in the east for all of El Salvador" (*una esperanza que nace en el Oriente para todo El Salvador*).

Like progressive Catholics, Anabaptists value social equality, mutual aid, and peaceableness, which they also perceive as characteristics of the reign of God, toward which Christians should look as a model for human society. In addition, Anabaptists also strive to make everyday experience efficacious in the transformation in and through voluntary religious communities. However, Anabaptists challenge many Catholics' close identification with particular political agendas. Because they insist on loyalty to only one ultimate authority, Anabaptists reject and withdraw from the corrupt secular world, in various degrees. The degree of separation varies widely, from liberal Mennonites to the Old Orders, although all agree that cooperation with the world must stop short of participation in the institutional violence of "national defense."

This distinctive kind of boundary-marking contributes to the development of common political goals in varied, sometimes ambiguous ways. The Amish seek not to affect politics but simply to be left alone in pursuit of their own idiosyncratic vision of religious redemption in the "believers' church." Amish involvement in politics is limited to local zoning ordinances or, in exceptional cases, to challenges to state or federal laws that threaten their religious lives, for example, to defend their right to educate children only until eighth grade. These legal struggles underline the fact that the Amish require certain concessions from the larger society in order to maintain their internal rules and ways of life. In general, however, the Amish hesitate to challenge the government; at the same time, they also avoid depending on it. However, a radical political potential lies in the Anabaptist understanding of community and of the community's relationship to the world. As noted earlier, a defining characteristic of Anabaptist thought is the insistence that communities live out their theology. The principal values that should be embodied in the believing community are those expressed in the Sermon on the Mount: discipleship, nonviolence, nonresistance (*Gelassenheit*), and mutual aid. By living out these values in community, according to Mennonite theologian John Howard Yoder, Anabaptist communities pose "an unavoidable challenge to the powers that be and the beginning of a new set of social alternatives."[4] The Christian community both separates itself from the world and presents a real alternative for the world. The expression of this alternative—its "promissory quality"—is at the same time its embodiment and realization. Through congregational decision-making and volunteer leadership, mutual aid and mutual corrections, Amish communities embody "a different way of being" in the here and now.

Relations between the Establishment of Political Goals and Efforts to Achieve Them

Once we understand some ways in which Anabaptists and progressive Catholics might develop political goals in and through their common boundaries and communal frameworks, we need to ask how these goals might be implemented in reality. What, if anything, works? This question preoccupies many students of progressive Catholicism, which, after generating great hopes for social transformation during the 1970s and 1980s, appears to have lost considerable momentum in the past decade or so. Progressive religious groups seem to face a crisis throughout Latin America, as Daniel Levine and David Stoll write:

> They have lost membership, political allies, and financial support. The resulting organizational decay has been deepened by indifference or hostility from former allies in church, society, and politics. Difficulties of this kind suggest that despite the fact that religious change has clearly empowered ordinary people throughout Latin America, there is a palpable gap between this 'empowerment' and 'power'—between the new energies and orientations spawned by religious change and the capacity of communities and organized groups to achieve tangible and durable benefits.[5]

Levine and Stoll ask about the links between personal empowerment and political power, on the one hand, and between experiments in local democracy and the transformation of larger political structures and institutions, on the other. Without denying that personal and local changes are significant, they suggest that they may not be significant in as straightforward a way as some activists and observers initially imagined.[6]

Religion plays a crucial role in building new cultural formations and transforming basic attitudes as well as institutions, in regards to political authority, economic justice, and human relations to nonhuman nature. Here both progressive Catholics and Anabaptist experiences are instructive: they begin with the "stable social groups," worldviews, day-to-day activities that Levine and Stoll identify as crucial. These religious movements undertake precisely the deep and long-term embodiment of principles in everyday lives and local communities that seems to be a prerequisite for major political and institutional changes. Communities need to define themselves and incarnate their values within their common borders before they can effectively carry their aspirations across those borders.

Here small Christian communities, whether Amish church districts or Roman Catholic *comunidades de base*, may play a special role. The experience of local democracy in the congregational model that both kinds of communities adopt shows an option between radical individualism (a charge Catholics often throw at Protestants) and unthinking collectivism (the Protestant countercharge). This alternative involves genuinely voluntary communities that "can affirm individual dignity (at the point of the uncoerced adherence of the member) without enshrining individualism."[7] Such communities also provide emotional, moral, and material support that enable members to continue living out their values when that would not be possible for them in isolation. No individual carries the burden of responsibility alone.

The progressive Catholic vision that has shaped peasant activism in El Salvador and elsewhere in Latin America since the late 1960s includes a "both-and" vision of the relationship between means and ends. Small communities such as CEBs and peasant cooperatives were seen as both the living embodiment of alternative values and a practical tool to implement these values in secular culture and political institutions. They were an end in themselves and at the same time a means to other ends. This makes sense in light of the long-standing Catholic insistence that Christians can and should embrace work "in the world" and that efforts for social justice, charity, or other religiously sanctioned goals have salvific, as well as moral, value.

The Amish show that it is possible to set up a community so that people live their values even when they conflict with larger values, cultural patterns and institutions. This is most evident in transportation. While most people in the United States could never get along without their cars, the Amish manage to very well because their institutions and communities make it possible for them to live out their values. This is a crucial lesson for building sustainable communities. Infrastructure, planned development, and local institutions (formal and informal) are necessary, as are knowledge of and cooperation from larger authorities. This is problematic in theory and in practice for Anabaptists, as for many local environmental activists who seek as much self-sufficiency and decentralization as possible.

The Salvadoran repopulations offer some different lessons and insights. Some of these are especially relevant for environmentalists who might not always keep these issues front and center. First, the history of the repopulations underlines the importance of institutions and political mobilization and organization in enabling communities to protect their own concrete interests and survival. The communities' political savvy, tenacity, and militance has made possible all their achievements, from early peasant organizations, throughout their experiences in the refugee camps, and continuing in each moment back in El Salvador. From the beginning, the Salvadoran government treated repopulations as military targets, although the residents asserted their rights as civilians. The movements have faced legal obstacles, political harassment and repression, and economic blockades. They have benefited, on the other hand, from international economic and technical aid and political support from NGOs and various organizations within El Salvador. The refugees' own organization and persistence, however, have ultimately been most momentous.

Tied to their ability to organize themselves politically, people in the repopulations have cultivated coalitions and strategic relationships at multiple levels, horizontally and vertically, to pursue shared goals and interests. Repopulations have established links with a host of entities, from local to international levels. Tactical alliances sometimes cannot overcome mutual suspicion and dissolve once specific objectives are achieved. However, on occasion deep and long-lasting relationships are forged in the process of working for a specific goal. While such coalition-building is less obvious among the Amish, it is worth noting that in a number of cases they have allied with environmentalists, preservationists, or nonfarming locals to preserve land or prevent an intrusive industry from settling in their areas.

Conclusions

Bioregionalism is an approach in theory and practice that understands human communities explicitly in relation to natural communities in local places. Wes Jackson has summarized bioregionalism's central claim nicely: "the majority of solutions to both global and local problems must take place at the level of the expanded tribe, what civilization calls community."[8] Thus, bioregionalist thinking and practice are essentially about communities, both cultural and natural. The aim is not to take people out of the land but rather to identify ways in which human settlements can understand, protect, and live in nondestructive relations with the natural world. Bioregionalism uses local communities as an ethical, epistemological, and political starting place: primary moral and emotional *attachment* is to the local place; most important kinds of *knowledge* concern local processes, history, inhabitants, species; and principal form of social *practice and activism* are locally grounded. But how do we define "local"? Here we return to the question of boundaries. Bioregionalism proposes a new way of drawing boundaries around human and natural communities, based not on government but on ecosystem and watershed, human and natural history.

Bioregionalism redraws boundaries, but also crosses them: between natural and cultural history, human and nonhuman creatures, local and global change, and more. Humans, like many other species, are "boundary creatures," as Michael Vincent McGinnis explains: "A boundary creature inhabits more than one world; the salmon, bear and people are linked and nested 'parts' of several distinct but interdependent systems of relationships."[9] Keeping these systems of relationships distinct yet interconnected is one of bioregionalism's chief aims. Bioregionalism not only crosses boundaries but also, perhaps paradoxically, makes them normative. It affirms the importance of setting limits and of seeing the borders of one's possible and appropriate spheres of action and attachment. In all this, we ought not to absolutize borders, but rather recognize that they are fluid, multilayered, overlapping, and permeable.

A local focus, then, carries dangers as well as advantages. Greater loyalty to a local place and its culture can lead to parochialism and intolerance toward differences or outsiders; or a focus on achieving local gains might lead to political ineffectiveness or indifference at larger levels. Some of these critiques have been raised in relation to the ascendance of "community-based" conservation and development in recent years. "Community-based" means various things, but usually it calls for taking into account the needs, interests, and perceptions of local communities in the development of conservation and development strategies. The definition of the community at stake, the kind and degree of interaction, and the community's influence on ultimate decisions all vary. Nonetheless, advocates of community-based models agree that conservation programs developed in isolation from or even opposition to local communities are less just or less effective, often both. Top-down models have been shown, again and again in different parts of the world, to fail to protect the environment or to achieve better qualities of life or greater justice for human communities. The kinds of projects that succeed more often than not resemble those in the Salvadoran repopulations, chosen, shaped, and managed by community members themselves and responding to concrete local needs and experiences.

Attention to the advantages of community risks glossing over the conflicts within communities, presenting them as more satisfying to members than they actually are, or ignoring changes within and around them. These dangers result from holding an idealized vision of community as stable, homogeneous, and harmonious, developed in the abstract and imposed on real communities without adequate attention to their real experiences. Another risk lies in thinking that local communities are all that matter in efforts at conservation or other kinds of positive social change. This can encourage researchers and practitioners to ignore what goes on at different levels, such as regional or national governments and markets, and also to underplay the interactions and relations among communities and between communities and other social forces. We should not think of small local communities, in short, as isolated from larger cultural, political, and economic processes.

"Community" is a normative as well as descriptive term. The writing on community-based conservation sometimes makes the mistake of applying a normative, idealized vision of community to fallible and idiosyncratic real communities. However, ideas about community are normative not only for scholars but also for community members themselves. Often they know that the community to which they belong fails to live up to its own ideals and self-image. Speaking of the community in idealized terms does not necessarily reflect ignorance or deceptiveness. Many times it reflects a hope and aspiration for what the community should be and could become, amid awareness that it currently falls short and perhaps always will.

Here we return to religion. Arran Garé describes people seeking to live according to ecological values as similar to the Roman Christians who St. Augustine described: "strangers in the societies in which they must live their everyday lives." Such groups, Garé writes, cannot simply withdraw into monasteries to wait for a new world, but must "begin to build this new civilization while the old civilization, despite its nihilism and the fragmentation of its culture, is still vigorous and powerful; more powerful than any civilization which has ever existed."[10] In their own way, the communities I have described strive to do just this. They present especially fruitful models for understanding the boundaries of the local and the global and the possibilities of border crossings. They do the same for the boundaries around the real and the utopian. Crossing those borders embodies the long-standing tension within Christianity between the "already" and the "not yet" of the reign of God. While the longed-for utopia will always be "not yet" on earth, it is already among us in these small and vulnerable communities, seeds and promissory notes for the future.

Notes

This chapter is based on work done for my book *Seeds of the Kingdom: Utopian Communities in the Americas*. New York and Oxford: Oxford University Press, 2005. I would like to thank Charles S. Brown and an anonymous reviewer for their helpful comments.

1. Donald B. Kraybill and Carl F. Bowman, *On the Backroad to Heaven: Old Order Hutterites, Mennonites, Amish and Brethren*. Baltimore: Johns Hopkins University Press, 2001, 105–106.

2. Most Amish try at least to work for Amish employers, to simplify their paychecks, since the Amish pay income taxes but, with Supreme Court permission, do not pay Social Security taxes. They argue that communities have the right and the obligation to take care of their own, and that accepting government insurance would undermine their mutual aid practices.

3. Holmes Rolston III, *Environmental Ethics: Duties to and Values in the Natural World*. Philadelphia: Temple University Press, 1988, 39.

4. John Howard Yoder, *The Politics of Jesus*, 2nd ed. Grand Rapids: William. B. Eerdmans., 1994, 39, 43. Yoder's phrase echoes the title of an essay by feminist poet Audre Lorde: "The master's tools will never dismantle the master's house."

5. Daniel Levine and David Stoll, "Bridging the Gap Between Empowerment and Power in Latin America," in *Transnational Religion and Fading States*, eds. Susanne Rudolph and James Piscatori. Boulder: Westview Press, 1997, 64–65.

6. Levine and Stoll, "Bridging the Gap," 66.

7. Yoder, *The Priestly Kingdom*, 24.

8. Jackson, *Becoming Native to this Place*, 2–3.

9. Michael Vincent McGinnis, "Boundary Creatures and Bounded Spaces," in Michael Vincent McGinnis, ed., *Bioregionalism*. London: Routledge, 1999, 61.

10. Arran Garé, *Postmodernism and the Environmental Crisis*. London: Routledge, 1995, 144.

References

Garé, Arran. 1995. *Postmodernism and the Environmental Crisis*. London: Routledge.

Jackson, Wes. 1994. *Becoming Native to this Place*. Lexington: University Press of Kentucky.

Kraybill, Donald B., and Carl F. Bowman. 2001. *On the Backroad to Heaven: Old Order Hutterites, Mennonites, Amish and Brethren*. Baltimore: Johns Hopkins University Press.

Levine, Daniel, and David Stoll. 1997. "Bridging the Gap Between Empowerment and Power in Latin America." In *Transnational Religion and Fading States*, eds. Susanne Rudolph and James Piscatori. Boulder: Westview Press.

McGinnis, Michael Vincent. 1999. "Boundary creatures and Bounded Spaces." In *Biogregionalism*, ed. Michael Vincent McGinnis. London: Routledge.

Rolston, Holmes III. 1988. *Environmental Ethics: Duties to and Values in the Natural World*. Philadelphia: Temple University Press.

Yoder, John Howard. 1984. *The Priestly Kingdom: Social Ethics as Gospel*. South Bend: University of Notre Dame Press.

———. 1994. *The Politics of Jesus*. 2nd ed. Grand Rapids: William B. Erdmans.

Chapter 10

The Moral Economy and Politics of Water in the Arid American West

T. Clay Arnold

On May 25, 1922, the *Inyo Register*, an Owens Valley, California, newspaper, proclaimed its support for the formation of a locally financed and administered irrigation district. Outraged over the City of Los Angeles's by then unmistakable (and ultimately successful) grab for Owens Valley water, a grab valley residents viewed as ruinous and unjust, the *Inyo Register* rallied its readers, writing: "Let us set aside all considerations of individual squabbles and relative rights and unite on the great issue of keeping for the Owens Valley, to be adjusted among ourselves, the water which by nature and justice belongs to it. Give the irrigation district your support. Any other attitude will be that of individualism rather than community; and in the end, what is best for the whole community is best for the individual interest" (Cited in Walton 1992, 166).

This remarkable proclamation resonates with the spirit of mutualism, community, and municipality. It illustrates the moral economy of water in the arid American West, that is, the fundamental, at times taken-for-granted normative principles that ultimately inform westerners' determinations of the legitimacy or illegitimacy of existing or proposed water-related practices, developments, or policies.[1] As revealed in incidents like the Owens Valley affair, as well as in historic and contemporary institutions and legislation, the moral economy of water in the arid West reflects the justice of communities ordering their own relations and welfare.

Interestingly, the moral economy of water often escapes the notice of many scholars of western water politics and policies, perhaps due to the methodological constraints of their preferred forms of inquiry. In these accounts, scholars typically explain outcomes in terms of elitist, pluralist, institutional, or market culture dynamics and imperatives. Among those who speak of a moral economy by name (notably Walton 1992), water plays only an incidental role at best.[2] A more accurate depiction of the moral economy of water, arguably the foundation for a more complete grasp of western water policies and history, rests initially on reconstructing "why people [in the arid West] *care* about water" (Bates et al. 1993, 178). The key to the interpretation that follows lies in recognizing with westerners water's status as a social good.

In one sense, all goods are social goods; every object or quality recognized as a good necessarily reflects culturally generated and culturally transmitted conceptions of human needs and benefits. As much as with the meanings and uses of words in a language, goods are the products of social, not individual or idiosyncratic, processes. Goods are social in effect as well as origin. Goods consist of shared understandings about their place and purpose in the course of human life. Individuals orient and coordinate their lives accordingly (Walzer 1983).

Some goods are social in another, deeper sense; they establish, reproduce, and/or symbolize important individual and collective senses of self. Possession and distribution of these goods define (a) people in particular and socially important ways, in some cases even to the point where the meaning of the good in question cannot be separated from the value of the identities and relationships generated.[3] Water in the arid American West is a case in point.

Westerners care about water for many reasons. Water provides an impressive array of agricultural, commercial, industrial, recreational, environmental, and aesthetic benefits (Wilkinson 1990), and the history of the West is marked by eras where one or more of these values assume a temporary priority (Hundley 1992; Pisani 1984, 1992; Worster 1985). These shifting regimes of value, however, are punctuated from time to time by another, more basic value—community or, alternatively, municipality.[4] Broadly speaking, water's relation to community in the arid West is twofold. First, water's historic role as the decisive material precondition for human settlement has endowed it with a special meaningfulness that includes the purposes and benefits of community. Second, water is a vital medium for social and political relations, a medium at the center of processes of community livelihood and self-identification.

Water is quite scarce in large parts of the West. Many areas average less than 12 inches of precipitation per year; some sections receive less than 5, far less than the 40 inches per year common in the eastern United States. No other physical feature has had as lasting an impact on people's efforts to live and flourish in what was once called the Great American Desert. Evidence of surprisingly sophisticated waterworks date to 300 B.C., lending credibility to claims that the West's persistent aridity has exerted a powerful social and political influence (Webb 1931, 17; Worster 1985, 7). Simply put, the search for, development, use, and control of water have been the very basis for settlement. At minimum, survival has been a matter of social and political adaptation, of adjusting the conduct of collective life to the imperatives of water acquisition and administration.

The physical realities of an arid environment affect over time how water is perceived and understood, at least for those ever on the edge of having to do without it. For example, native and Spanish inhabitants of much of what was to become the American Southwest matter-of-factly regarded water as the "prime determinant" of their individual and collective lives. Not surprisingly, they readily afforded water "a telling reverence" in their religion, mythology, and lore, a cultural imprint that lingers to this day, particularly in the Upper Rio Grande regions of New Mexico and Colorado.[5] Modern residents of the arid West revere water in their own way, regularly characterizing water as life-giving, lifeblood, precious, and fundamental (Mumme and Ingram 1987).

Historically and culturally, residents of the arid West have also perceived and understood water from the perspective of the promise of community. The two are inextricably intertwined. Effective and lasting mechanisms for acquiring and distributing water go well beyond the abilities of even the most able individual. Water development and management entail a coordinated, collective endeavor rooted in the frank recognition of mutual dependency and joint responsibility (Sax 1990, 17). Successful acquisition and administration are in turn the basis for the associated benefits of "schools, churches, and social life" (Mead 1903, 382), for a civic and moral as well as economic progress. Then, as now, westerners understand water from the municipal perspective of inherently desirable states of affairs achieved and experienced only in concert with others.[6]

This understanding of water has and continues to influence western water policies and practices, and is perhaps most clear in the *acequia* (irrigation) communities once found throughout the greater Southwest but today concentrated in New Mexico and Colorado (Rivera 1998). For Spanish colonizers of the arid Southwest, the multiple and clustered benefits of human settlement turned on the establishment of a municipal order where communal principles were simultaneously the means for and products of provision. The municipal order they created, famously codified in the *Plan de Pitic* (1783–1789), was that of the community ditch or *acequia madre*. Figuratively if not literally at the center of the settlement, the ditch was community property and as such "was subject to regulation by municipal officials," including "construction, maintenance and repair, and distribution of its waters." For any given member of the *acequia*, "the community assured him the right of common use and at the same time imposed upon him responsibilities for assisting in the upkeep and conforming to the rules governing water use" (Clark 1987, 15). Present day *acequias* still employ, indeed, celebrate these now traditional communal ways and means (Crawford 1988; Rivera 1998).

Mormon settlers of the American West also recognized the municipal nature of water. "There shall be," Brigham Young pronounced, "no private ownership of the streams that come out of the canyons" (Dunbar 1983, 13). County courts (boards consisting of a probate judge and three selectmen) were charged with regulating the water, and in such a way as to best meet "the interest of the settlements in the distribution of water for irrigation" (14).[7]

Later Anglo irrigation practices were far more developmental, extractive, and, according to many, market-driven. Yet here, too, settlers understood water as "the basis for economic growth that supported learning, religion, and community" (Sherow 1990, 1). James Sherow describes these settlers' animating vision as one of "cooperation and prosperity premised upon the mutual and public ownership of water" (1). Their preferred institution was the mutual irrigation company, or cooperative, which, according to Dunbar (1983, 28), "grew out of the neighborhood or community construction of ditches." Typically founded by farmers, mutual companies were "democratically regulated by granting shareholders a division of canal flow, assessing them for upkeep, and allowing them a vote in policy formation in proportion to their holdings" (Sherow 1990, 12). Early champions of the mutual irrigation company, individuals like Theodore Heinz and William Smythe, emphasized that, in the arid West, water was "too important to be left in the hands of individuals" (12). Contemporary water laws and policies, broadly and consistently regulatory, also reflect the public, municipal nature and understanding of water. The exchange, sale, appropriation, or transfer of water in the arid West operates within the bounds of water's public resource status, in particular western states' public interest and especially public trust obligations (Ingram and Oggins 1992; Sax 1989).

Although the settlement era is long past, water still serves as the basis for association and joint endeavor. Throughout the 1990s, for example, westerners organized upwards of 800 watershed associations, councils, or partnerships for addressing a wide range of water-related concerns. These associations enjoy the participation and official support of both federal and state agencies.[8] Described by Matthew McKinney, director of the Montana Consensus Council, as the "rejuvenation of Jeffersonian democracy" (1998, 33), collaborative watershed partnerships emphasize shared interests and facilitate the pursuit of a common good. Scholars and advocates alike have highlighted the *"intrinsic value"* of these associations' collaborative arrangements, that is, an enhanced "*sense of community and place, which in turn, improves the quality of life for all residents, and improves the ability of a community to achieve social, economic, and environmental goals*" (Kenney 2000, vi; italics in the original). Secretary of the Interior Bruce Babbitt's assessment of the watershed movement makes a similar point. Describing the movement as "one of the most refreshing, encouraging trends on the entire American landscape," Babbitt concludes that "Once people are in there doing it with their own two hands... it gives them a sense of [a] kind of ownership of the result and a lot more intensive feeling about the importance of it and the need to really put it all together." Watershed partnerships reflect the "idea that local citizens get together and have an organizing principle which is not, you know, just a particular 40-acre tract or something over or behind the neighbor's fence. And the waters literally bring them together" (2000, 2, 3).

Given water's scarcity and importance, individuals, communities, and regions in the West cannot help interacting with one another on the basis of shared or contested rivers, watersheds, or aquifers. More to the point, where a given individual, community, or state falls within the extensive network of physical, legal, and administrative institutions that capture and allocate water (1) defines them as one kind of agent or another (e.g., water-rights holder, junior appropriator, downstream community, upper-basin state, etc.); (2) establishes to a point any given agent's legal, socioeconomic, and political capacities; and (3) vests agents with certain kinds of interests and predispositions.[9] Water relationships directly affect how states and communities interpret and pursue their respective interests and, in the process, assert or maintain their identities as states and communities (Espeland 1998, xii). The moral economy of water is intimately bound up with the municipal integrity of states and communities.

Arid western states and communities understand that their status as viable and rewarding social, economic, and political associations rests precariously on the supply and disposition of water. As one community leader put it, when faced with the prospect of a sharp reduction in the supply (and especially control over) water, "the community loses its present, its past, and its future" (Ingram and Oggins 1990, 5). The fear of suffering just such a fate, especially at the hands of a rival, explains westerners' historic preferences for municipal control and their opposition to water-related policies that would, in their estimation, diminish a political association's ability to order its own relations and welfare, thereby compromising its identity as a community. California's passage of the Wright Act (1887), and its subsequent adoption in sixteen other arid and semi-arid states, illustrates this.

Penned by C. C. Wright, the act created the irrigation district, an entirely new form of public authority expressly for the purpose of fostering and securing community-based irrigation. Radically local in nature, the act permitted residents to form, with a two-thirds vote in an open election, a special district government vested with considerable powers, among them the power to tax property and issue bonds; purchase and condemn water rights and rights-of-way; construct dams and canals; and distribute the water supply (Pisani 1984, 253). The Wright Act was a conscious response to the growing threat of water monopoly. Californians in particular had reason to worry. The preexisting doctrines of riparianism and prior appropriation did not and could not support over time the practices and expectations of thousands of small landholders and hundreds of communities whose futures turned on widespread irrigation.[10] As applied in California, riparianism and prior appropriation increasingly confined the use and benefits of water to a narrow and thereby privileged few, denying large and growing numbers of farmers and associated communities ready and certain access to life-sustaining water. Frustration and opposition mounted. Delegates to an irrigation convention two years prior to the Wright Act put the matter quite plainly. Speaking of the monopolistic nature of riparianism in particular, they asked, "Which is better? That a few men, the limited few who own the [river] bank, should have the exclusive use of the stream to water their stock, all irrigation be stopped, all progress of the past blotted out, and ruin and destruction to all the prosperous and happy homes of which now irrigation is the cause, the life and only hope; or that the stream be used so as to irrigate the greatest amount of land which it is capable of irrigating, so as to stimulate production to its widest limits, so as to build up homes of plenty and happy firesides and rich and prosperous communities and peoples?" (Cited in Hundley 1992, 90–91).

The monopoly Californians feared was not simply an economic one, although many were certain that, if not checked, economic monopolies would quickly emerge, with water barons in effect extracting undue financial advantage from the great body of the people (Alston 1978). Significantly, the specter of monopoly also included a pronounced sense of undue social and political dominion, with water in this case allowing water barons "to monopolize not [just] business, but," as the *Sacramento Bee* put it in 1874, "the main elements of all life" (Pisani 1984, 117). For Californians, and those who later adopted the principles of the Wright Act as their own, the moral economy of water demanded an official state of affairs much more attuned to water's role as a social good, to a people's shared future as a people with water in an arid land. Unlike riparianism (imported from the water-rich East) or prior appropriation (initially devised to meet the needs of itinerant miners in the 1850s), Wright's emphasis on community-based irrigation districts directly addressed and codified what Californians had come to understand about water but struggled to secure. The Wright Act passed unanimously.

Contemporary westerners are no less sensitive than nineteenth-century Californians to the role of water in securing or seriously damaging the autonomy and integrity of states and communities, especially given the West's current focus on efficiently managing the redistribution of its more or less fully developed supply of water. Western states and communities have responded in ways that confirm the moral economy of water.

Consider the example of the City of El Paso, Texas, and its efforts throughout the 1980s to compel the state of New Mexico to allow El Paso to drill and operate over 300 groundwater wells in the Mesilla aquifer of southcentral New Mexico. Citing the U.S. Constitution's interstate commerce clause, El Paso filed suit in federal court to overcome New Mexico's longstanding ban on exporting water. The court found in favor of El Paso, declaring New Mexico's blanket ban on exports an unconstitutional economic advantage. New Mexicans viewed El Paso's efforts as nothing less than a direct threat to their integrity as a state, as one official put it, "a well-planned invasion by a foreign country," an invasion designed "to keep us like a colony" (Barrilleaux 1984, 22, 42). Editorials spoke of New Mexico's border as having been "breached" and threatening "the fundamental authority of the state."[11] Ira Clark, resident and noted historian of New Mexico, agreed, concluding in an interview, "El Paso is only the tip of the iceberg. If El Paso succeeds, people all around us—Denver, Phoenix, Tucson, Amarillo—are waiting to take water out of New Mexico, too. We're very worried about that."[12] Individuals, districts, cities, and counties throughout the affected region jointly mobilized to contest El Paso. Citizens of Las Cruces, New Mexico, organized an economic boycott of El Paso and distributed "Thou shalt not covet they neighbor's water" bumper stickers. The 5,000 residents of Chaparral, New Mexico, raised $47,800 through rummage sales and dances to cover their share of the legal costs (Harris 1990, 23). The City of Alamogordo, New Mexico, three counties, the State Land Office, and New Mexico State University participated as well. So, too, the state legislature; taking a cue from the court's decision, they passed legislation, subsequently ruled constitutional, forbidding interstate sales and transfers of groundwater not deemed in the "public welfare" of New Mexico, greatly complicating El Paso's efforts (Clark 1987, 674–676; Harris 1990, 17–18).[13]

Consider as well parallel developments at the very local level in Arizona and Colorado. As much as New Mexico in the El Paso case, rural regions and communities are today the target of growing metropolises seeking additional supplies of water, a situation very similar to that of the Owens Valley in the 1920s. Concerns about civic decline in La Paz County, Arizona, whose groundwater was sought by the cities of Phoenix and Scottsdale, prompted appeals before the state legislature and considerable debate over the justice of intrastate water marketing (price-driven rural-to-urban transfers) (Checchio 1988). In hearings before the Arizona Senate, residents of La Paz County characterized such transfers as "the 'rape' of the county's future," as unjust as "taking food from a small child."[14] Bruce Babbitt, former Governor of Arizona, described these kinds of situations as "economic Darwinism," a direct threat to "the very existence of rural communities that happen to be well-endowed with water." Invoking the moral economy of water, Babbitt urged greater protective regulations, otherwise, he noted, "Big cities will inexorably squeeze the water and the life out of small communities" (Babbitt 1988, 7).[15] The legislature responded, passing legislation containing the injunction: "Groundwater may not be transported away from a groundwater basin" (AZ Stat. 45-544).

An even more recent and revealing example is that of the San Luis Valley in south-central Colorado. Roughly 50 miles long, 75 miles wide, and receiving less than 10 inches of precipitation a year, the San Luis Valley is one of the world's largest high desert valleys. The Valley is also rural and agricultural; its 40,000 plus residents live in six of Colorado's least wealthy counties. Although arid, the valley sits atop a 6,000-foot deep aquifer containing tremendous amounts of groundwater (approximately two billion acre-feet). In 1986, a consortium of investors, led by Canadians and known as American Water Development, Incorporated (AWDI), applied for rights to 200,000 acre-feet of groundwater from the valley's aquifer. AWDI planned on selling that recovered water to thirsty western metropolises for very lucrative returns (anywhere from $4,000 to $7,000 an acre-foot).

Valley residents vigorously mobilized to defeat the proposal. According to Judge Robert W. Ogburn, who presided over the court case contesting AWDI's petition, valley residents "came together almost as one in their opposition.... They perceived the AWDI organization as an avaricious corporation that would suck the lifeblood of the Valley out of the ground and turn their verdant valley into another wasteland like the Owens Valley, California" (1996, 30). An "extraordinary coalition" (Bingham 1996, 139) of whites, Indians, Hispanics, farmers, ranchers, and environmentalists formed an entirely new political organization, Citizens for San Luis Valley Water (CSLVW). As a grassroots organization, CSLVW raised funds, published newsletters, sponsored public forums, organized rallies, and commissioned a large mural, displayed throughout the valley, colorfully depicting the valley's vital ties to its water. The mural also featured the inscription, "We the people of the San Luis Valley pledge to nurture and protect our resources and to promote justice and human dignity." Along with other valley organizations, CSLVW supported the legal battle against AWDI. Valley residents capitalized on the revenue abilities of the Rio Grande Water Conservation District. They proposed raising property taxes through the district to cover court costs, a particularly clear case of municipality for the sake of community. In a high-turnout election, residents of the poorest counties in Colorado voted 20–1 in favor of the tax increase (Bingham 1996, 139). AWDI's application was denied.[16]

Water's status as a social good is the basis for its moral economy, the framework for determining what makes policy "right" or "wrong." According to Michael Walzer (1983, 8–9), social goods generate their own normative principles. When we understand what a social good like water "means to those for whom it is a good, we understand how, by whom, and for what reasons it ought to be distributed. All distributions are just or unjust relative to the social meanings of the goods at stake." Although keenly aware of water's many uses and benefits, westerners ultimately understand water in communal and municipal terms, that is, in terms of often long-standing (or ardently sought) states of affairs, relationships, and identities embedded in water-related interdependence and association. Intrinsically communal and municipal meanings also derive from the mutually reinforcing situation of arid states and communities constantly striving to secure the water on which their futures as states and communities depend, an ever-evolving exercise in provision for the sake of community, community for the sake of provision (Walzer 1983, 64–67).

Expanding the supply of water through development is one way states and communities secure their status as effective civic associations. However, in arid environs, long-term collective prospects also rise or fall given the ability of states and communities to govern in some sense the distribution of water. Given the intrinsically communal properties of water, arid states and communities have a compelling and legitimate claim to the disposition of water, at least to the point of jurisdiction over forms of distribution (e.g., fully self-interested, privatized, and/or market-driven transactions) that might otherwise deprive states and communities of the ability (and justice) of ordering their collective affairs and welfare. The moral economy of water is in this sense negative, historically opposed to practices and policies that do not take water's status as a social good appropriately into account.

This moral economy of water accounts for westerners' long-running resistance to treating water merely as a commodity, a requirement at the philosophical heart of most proposals for more private and price-driven markets in water. According to conventional economic thought, commodities are the "instruments of commerce and circulation" (Smith 1973, 72), that is, discrete objects whose most significant feature is that they possess exchange value. So conceived, the value of any commodity is what it brings in a commercial transaction. Commercial markets in turn presume the commensurability of all commodities, with their commensuration achieved through the medium of money. Finally, exchange is entrepreneurial (i.e., individualistic and utilitarian), centered on extracting net financial gain. Yet, for westerners, water is far more complex than the commodity view allows. It is a social good in the deepest sense of the term. Moreover, as the "lifeblood" of community, westerners recognize in water a value beyond that which is realized in commercial exchange, indeed, a value they have long considered as not properly for sale.[17]

Reduced to its most elemental form, the moral economy of water reflects westerners' deep concerns about water-related practices and policies that expose states and communities to external, arbitrary control, compromising their futures and identities as autonomous and rewarding associations. The challenge before us today, much like it was for residents of the Owens Valley in 1924, consists in recasting water's moral economy in far more affirmative principles and prescriptions, a task rooted in grasping even more thoroughly water's intrinsically communal properties.

Notes

1. On the general idea of a moral economy, see Arnold 2001, Booth 1994.

2. Walton's study is of the Owens Valley affair, of rebellion rooted in outrage over violations of the residents' moral economy. In Walton's interpretation, however, the moral economy residents sought to uphold was not centered on water but the legacy of a pioneering tradition wedded to commercial economy and as informed by Progressivism, or, as Walton puts it, the "cultural premises of paternalism, manifest destiny, and the laissez-faire economy" (1992, 328). Feige's (1999, 252, n. 59) one-sentence reference to moral economy concerns the cooperative nature of late nineteenth- and early twentieth-century irrigation practices in Idaho, which he describes simply as "an outgrowth of the social relationships among people of the upper Snake River valley."

An exception to the literature on western water policies, one to which this chapter is indebted, is the work of Ingram and several co-authors (Brown and Ingram 1987; Ingram 1990, 1992; Ingram and Oggins 1990; Mumme and Ingram 1985; Nunn and Ingram 1988). Although these works do not employ or mention the concept of a moral economy, their focus on the noncommodity and community-oriented nature of water highlights much of what I refer to as the moral economy of water.

3. Ethnography and cultural anthropology have demonstrated that knowledge of systems of status, prestige, identity, kinship, power, and the like is often best gleaned from careful attention to the "cultural biographies" of various socially esteemed objects, be they plots of land, cattle, rice, tobacco, pots, or brass rods (Kopytoff 1986). These highly esteemed goods generally "provide the necessary conditions for entry to high-status positions, for maintaining rank, or for combining attacks on status" (Appadurai 1986, 25). Insofar as these goods establish, reproduce, and/or symbolize relations between people, they directly affect the social, economic, or political capacities of a person or a people.

4. I use these two terms heuristically, to underscore two different, though often overlapping aspects. In its strongest form, *community* refers to the sense of collective purpose and identity brought about by underlying affective ties and circumstances.

Municipality refers more to the institutions or mechanisms of attending to a community's collective or public affairs. The distinction is not absolute; although community need not lead to municipality, it often does. Moreover, municipal institutions and mechanisms may just as easily reinforce or reproduce community.

5. Meyer (1989, 4, 76–78, 101) emphasizes the distinctive worldview, value structures, and thought processes of those involved in the incessant struggle with aridity in the American Southwest.

6. Water in the arid West illustrates Sagoff's (1986, 302, 315) point about public values: "Public values are goals or intentions the individual ascribes to the group or community of which he is a member; they are his because he believes and argues they should be ours; he pursues them not as an individual but as one of us. The individual then shares with other members of his community *intersubjective* intentions or, to speak roughly, common goals and aspirations, and it is by virtue of these that a group or community *is* a group or community.... [P]eople in communities know purposes and aspirations together they could not know alone."

7. See Maass and Anderson (1978) and Fiege (1999) for how these sentiments affected twentieth-century practices in Utah and Idaho.

8. In 1996, the Environmental Protection Agency (EPA) concluded: "These watershed approaches are likely to result in significant restoration, maintenance and protection of water resources in the United States. Supporting them is a high priority of EPA's national water program" (in Kenney 2000, 4).

9. The very word *rival* comes from the Latin term *rivalis*, referring to those who live on opposite banks of a stream (Maass and Anderson 1978, 2).

10. Riparianism grants water rights only to those whose property borders the stream. Riparianism guarantees all riparian landowners an unimpeded, natural, and fully flowing watercourse. Riparians may take only that amount of water necessary for stock and everyday domestic uses. Riparianism does not authorize withdrawals for irrigated agriculture. Prior appropriation awards rights to those first to claim, capture, divert, and put to "beneficial use" flowing water ("First in time, first in right"). Appropriation rights last only as long as actual diversion and use ("Use it or lose it"). Early or senior appropriation rights preempt all later or junior claims, a distinct advantage when flows fluctuate, as they often do in arid areas, or when later claims exceed available supply.

11. *Albuquerque Journal*, January 21, 1983, A-4.

12. *Dallas Times Herald*, September 12, 1988.

13. Compare with Arizona, which does permit out-of-state transfers but only after the director of the Department of Water Resources has taken into account, among other things, "Potential harm to the public welfare of the citizens of this state" (AZ Rev. Stat. 45–292).

14. *Phoenix New Times*, January 30, 1991, 99.

15. Communities and regions constrained to export water can decline in a number of ways. These declines include lost tax base, reductions in economic, public, and social services, population loses, and increases in the levels of citizen disengagement, malaise, defeatism, and fatalism. Nunn and Ingram (1988, 475) include the "redistribution of political authority over resource use from the area of origin to the importing region." Albert and Ruth C. Shaffer (1984, 310) emphasize reductions in a community's "adaptive capacity" (including the loss of leaders able to look and plan beyond the here and now), which can reach the point of compromising the rural community's historic role of "linking residents to basic values and social institutions."

16. A similar application, filed several years later, prompted valley residents to publish a full-page declaration in Sunday papers in Denver, stating: "We, the undersigned citizens, are committed to preserving the water within Colorado's San Luis Valley for land and life. Private marketers are drawing up plans to take water from the Valley and sell to other parts of Colorado or other regions. *We are opposed to these plans....* We are committed to fight efforts to take water from this beautiful place, the San Luis Valley. Help us protect land and life" (*Denver Post*, October 20, 1996, 5 B).

17. Brent Haddad (2000) attributes California's failure to establish a market for water, despite two decades of effort, to the extreme difficulty of creating a market mechanism that avoids the prospect of the highest bidders for water dominating other interests or regions. "The Ghost of Owens Valley," he concludes, hovers "over water planning, not just in California, but throughout the arid West" (xv, xvi). Compare with F. Lee Brown's assessment, made in response to proposals for instituting a water bank in New Mexico: "While we do have active markets **for water rights** in New Mexico, we don't truly have markets **for water itself**.... The quiet evolution of markets for water rights has obscured [an] underlying distrust of the marketplace's treatment of water purely as a commodity like any other and may have lulled some of us into the presumption that water itself could also be as readily bought and sold as are water rights" (2000, 1; emphasis in the original).

References

Alston, Richard M. 1978. *Commercial Irrigation Enterprise, the Fear of Monopoly, and the Genesis of Market Distortion in the Nineteenth Century American West*. New York: Arno.

Appadurai, Arjun, ed. 1986. *The Social Life of Things*. Cambridge: Cambridge University Press.

Arnold, Thomas Clay. 2001. "Rethinking Moral Economy." *American Political Science Review* 95 (March): 85–95.

Babbitt, Bruce. 1988. "Urbanized West Needs New Water Laws." *Los Angeles Times*, August 9, 1988, Part Two, p. 7.

———. Transcript of broadcast radio interview. Morning Edition, National Public Radio, June 28, 2000.

Barrilleaux, Ryan J. 1984. *The Politics of Southwestern Water*. El Paso: Texas Western Press.

Bates, Sarah, David H. Getches, Lawrence J. MacDonnell, and Charles F. Wilkinson. 1993. *Searching Out the Headwaters: Change and Rediscovery in Western Water Policy*. Washington, DC: Island Press.

Bingham, Sam. 1996. *The Last Ranch: A Colorado Community and the Coming Desert*. New York: Harcourt Brace.

Booth, William James. 1994. "On the Ideal of a Moral Economy." *American Political Science Review* 87 (December): 949–954.

Brown, F. Lee. 2000. "Do We Need Water Markets? YES, BUT...". New Mexico Water Resources Research Institute Report No. 312. Las Cruces: New Mexico State University.

Brown, F. Lee, and Helen Ingram. 1987. *Water and Poverty in the Southwest*. Tucson: University of Arizona Press.

Checchio, Elizabeth. 1988. "Water Farming: The Promise and Problems of Water Transfers in Arizona." Issues Paper No. 4, Water Resources Research Center. Tucson: University of Arizona Press.

Clark, Ira G. 1987. *Water in New Mexico: A History of Its Management and Use*. Albuquerque: University of New Mexico Press.

Crawford, Stanley. 1988. *Mayordomo: Chronicle of an Acequia in Northern New Mexico*. Albuquerque: University of New Mexico Press.

Dunbar, Robert G. 1983. *Forging New Rights in Western States*. Lincoln: University of Nebraska Press.

Espeland, Wendy. 1998. *The Struggle for Water: Politics, Rationality, and Identity in the American Southwest*. Chicago: University of Chicago Press.

Fiege, Mark. 1999. *Irrigated Eden: The Making of an Agricultural Landscape in the American West*. Seattle: University of Washington Press.

Haddad, Brent M. 2000. *Rivers of Gold: Designing Markets to Allocate Water in California*. Washington, D.C.: Island.

Harris, Linda. 1990. *Whose Water Is It, Anyway? Anatomy of the water battle between El Paso, Texas and New Mexico*. Las Cruces, NM: Arroyo Press.

Hundley, Norris. 1992. *The Great Thirst: California and Water, 1770s-1990s*. Berkeley: University of California Press.

Ingram, Helen. 1990. *Water Politics: Continuity and Change*. Albuquerque: University of New Mexico Press.

———. 1992. "Politics, Markets, Society, and Water Resources." *Halcyon* 14: 57–72.

Ingram, Helen, and Cy Oggins. 1990. "Water, the Community, and Markets in the West." Western Water Policy Project Discussion Series Paper, No. 6. Natural Resources Law Center, University of Colorado School of Law.

Kelley, Robert. 1989. *Battling the Inland Sea: American Political Culture, Public Policy, and the Sacramento Valley, 1850–1986*. Berkeley: University of California Press.

Kenney, Douglas S. 2000. *Arguing About Consensus: Examining the Case Against Western Watershed Initiatives and Other Collaborative Groups Active in Natural Resources Management*. Natural Resources Law Center, University of Colorado School of Law.

Kopytoff, Igor. 1986. "The Cultural Biography of Things." In *The Social Life of Things*, ed. Arjun Appadurai. Cambridge: Cambridge University Press.

Maass, Arthur, and Raymond L. Anderson. 1978. *...and the Desert Shall Rejoice: Conflict, Growth, and Justice in Arid Environments*. Cambridge: MIT Press.

McKinney, Matthew. 1998. "Working to Save Water in the West." *New York Times*, Late Edition (East Coast), June 14, 1998. Section 1, p. 33.

Mead, Elwood. 1903. *Irrigation Institutions*. New York: Macmillan.

Meyer, Michael. 1989. *Water in the Hispanic Southwest: A Social and Legal History, 1550–1850*. Tucson: University of Arizona Press.

Mumme, Steve, and Helen Ingram. 1985. "Community Values in Southwest Water Management." *Policy Studies Review* 5 (November): 365–381.

Nunn, Susan C., and Helen Ingram. 1988. "Information, the Decision Forum, and Third-Party Effects in Water Transfers." *Water Resources Research* 24 (April): 473–480.

Ogburn, Robert W. 1996. "A History of the Development of San Luis Valley Water." *The San Luis Valley Historian* 28: 5–40.

Pisani, Donald. 1984. *From Family Farm to Agribusiness: The Irrigation Crusade in California and the West, 1850–1931*. Berkeley: University of California Press.

———. 1992. *To Reclaim a Divided West: Water, Law, and Public Policy, 1848–1902*. Albuquerque: University of New Mexico Press.

Rivera, Jose A. 1998. *Acequia Culture: Water, Land, and Community in the Southwest*. Albuquerque: University of New Mexico Press.

Sagoff, Mark. 1986. "Values and Preferences." *Ethics* 96 (January): 301–316.

Sax, Joseph. 1989. "The Limits of Private Rights in Public Waters." *Environmental Law* 19 (Spring): 473–483.

———. 1990. "The Constitution, Property Rights and the Future of Water Law." Western Water Policy Project Discussion Series Paper, No. 2. Natural Resources Law Center, University of Colorado School of Law.

Shaffer, Albert, and Ruth C. 1984. "Social Impacts on Rural Communities." In *Water Scarcity*, ed. Ernest A. Engelbert and Ann Foley. Berkeley: University of California Press.

Sherow, James. 1990. *Watering the Valley: Development Along the High Plains Arkansas River, 1870–1950*. Lawrence: University Press of Kansas.

Smith, Adam. 1973. *The Wealth of Nations*. In *Masterworks of Economics*, vol. 1, ed. Leonard D. Abbott. New York: McGraw-Hill.

Walton, John. 1992. *Western Times and Water Wars: State, Culture, & Rebellion in California*. Berkeley: University of California Press.

Walzer, Michael. 1983. *Spheres of Justice*. New York: Basic.

Webb, Walter Prescott. 1931. *The Great Plains*. New York: Grossett and Dunlap.

Wilkinson, Charles F. 1990. "Values and Western Water: A History of the Dominant Ideas." Western Water Policy Project Discussion Series Paper, No. 1. Natural Resources Law Center, University of Colorado School of Law.

Worster, Donald. 1985. *Rivers of Empire: Water, Aridity, and the Growth of the American West*. New York: Pantheon.

Chapter 11

The Need for a Taxonomy of Boundaries

Wes Jackson and Jerry Glover

The Problem: The lack of a formal language adequate to bridge the middle ground between global accountants and bookkeepers of business.

We humans, as creatures of the Upper Paleolithic, have been around with the big brain, the 1,350 cubic centimeter brain, for some 150,000–200,000 years. During most of that time and before, we could take without thought for the morrow as our surroundings either met our daily needs or they did not. We could migrate or go to war, or both; become sick or die, or both. Nature more or less took care of itself, so there was little need to inventory the loss of ecological capital or standing stock. The global commons interacting with various locales stood behind sustenance and health.

Even during our early hunting and gathering periods, options for future generations at regional, if not at the global, level were squandered. The widespread extinction of large land mammals due to overkill by paleo-hunters during the Pleistocene and much later repeated occurrences of land degradation and resultant soil loss by ancient Greek farmers are but two examples of reduced ecological capital as stock.[1] The consequences of these past errors pale, however, when compared with the likely consequences of modern human-caused loss and despoliation.

Humans have appropriated nearly 40 percent of the earth's net primary productivity to date[2] and, given current trends, will claim a majority of the planet's biotic production as our own within a few decades. Human management of this biotic production relies largely on simplifying the complex, highly evolved, and self-regulating systems all fueled by the sun. These biotic systems have evolved over millennia and feature intricate feedback loops.

Due to the complexity of the systems, humans necessarily simplify the planet's biotic systems by replacing the regulators with their own complicated engineering and regulatory activities that feature ever-widening feedback loops over space and time. Humans, with only a several-hundred-year-old understanding, are attempting to manage several-hundred-millennia-old systems. Again, the failures of paleo-hunters and ancient Greeks will pale in comparison to the failure of modern humans if our understanding proves insufficient to manage fifty percent of this planet's biotic productivity.

A diagram of the complicated engineering and regulatory activities necessary to manage these complex systems resembles a near comical Rube Goldberg assembly that works well on paper. The problem is not simply that things in the real world go wrong, but that, because of the long feedback loops featured in human-designed systems, we are slow to realize the consequences. This is why the majority of modern accountants have so proudly proclaimed 8,000 years or so of deficit spending as economic growth.

Proclamations of economic growth occur despite increases in greenhouse gases, formation of a low-oxygen "dead zone" in the Gulf of Mexico from agricultural runoff, lowering of aquifers such as the Ogallala by irrigation, and the loss of nonrenewable topsoil. Conventional business accountants are not trained for such bookkeeping. Simple financial transactions reveal no trail showing all inputs, flows, transformations, and outputs. Entities and processes not relevant to efficient distribution of goods to the immediate human stakeholders are externalized by being ignored.

These externalized effects, which began to accelerate in the 1800s, have given rise to a new type of accountant, the forensic scientist, who has emerged in such fields as climatology, biology, and ecology. These new accountants tally the large-scale effects of our human business transactions and for some measures they can be quite specific. Concentrations of atmospheric carbon dioxide have increased from preindustrial levels of 278 parts per million by volume (ppm) to more than 350 ppm today.[3] Oxygen concentrations in the Gulf of Mexico can plunge from a typical level of seven milligrams per liter of seawater to less than two milligrams per liter of seawater.[4] Other global accountants have attempted to balance the whole of human economic activity by comparing the $25 trillion value of annual world economic product against the estimated $33 trillion value of "nature's services"—a rough estimate of the value of the natural processes necessary to conduct all business transactions.[5] If this means that the value of the "capital" necessary to do business on the planet is $33 trillion, yielding a product of $25 trillion, what is the depreciation rate of the capital?[6]

Some of these global accountants have gone beyond scorekeeping and have begun sniffing back up the trail to the source. Rachel Carson is an early exemplar of this type of investigative global accountant. More contemporary examples include researchers at Louisiana Universities Marine Consortium who traced the cause of the hypoxic zone in the Gulf of Mexico upstream to the industrialized farms and animal confinement operations of the Midwest.

As we approach the fifty percent mark of human appropriation of the earth's biotic productivity, we must move beyond these examples of global accountants that work back upstream. We need accountants whose work is devoted to warning us away from situations requiring expensive solutions or, worse, situations with no way out. Richard's desperate plea, "A horse! A horse! My kingdom for a horse!" as he finds himself being surrounded in Shakespeare's *Richard III* fits the problem toward which we are rushing.

Wendell Berry wrote the first author of this chapter that "when we came across the continent cutting the forests and plowing the prairies, we have never known what we are doing because we have never known what we are undoing." This statement acknowledges the failure of our current accounting systems from the time Europeans "opened" the continent. We need a *preemptive* accounting system to close the disciplinary gap between the global accountants and the bookkeepers of business.

Preemptive accounting, featuring middle ground accountants, will require a science of boundaries where the most important effects are measured: the difference between oxygen concentrations in a dead zone and normal seawater; the difference between topsoil depth under native prairie vegetation and under annual cropping; the difference between benefits and costs. The modern botanical taxonomic system revolutionized our understanding of the relatedness of the living world. Taxonomists keep the record straight on similarities and differences among life forms and have greatly increased our understanding of their form and function. The taxonomic system itself provides the necessary formal language in which biological relationships are discussed and accounted for. A boundary taxonomy could similarly advance the accounting of human activities.

Proposed Solutions That Are Inadequate

Neoclassical Economics

Not all "middle ground" activity (between the global accountants and the bookkeepers of business) can be adequately analyzed, so we are left to the machinations of the market where the dominant idea is the neoclassical economic model. The seat of this model rests on three legs: the market, education, and public policy. The model works when knowledge is adequate and political will is sufficient to draft, pass, and enforce legislation. Then the market can work if the ecosystem has not gone beyond recovery.

The neoclassical model can solve the dead zone problem in the Gulf of Mexico—*theoretically*—because oxygen deficiency is largely the consequence of too much nitrogen washed seaward from the farm fields of the Mississippi River basin. A reasonable estimate of the amount of nitrogen responsible for the growth leading to hypoxia is possible. From there on we can determine what level of nitrogen could properly be applied across the watershed and then through policy enforce the necessary reduction. There are several essential steps, but the problem could be solved—*theoretically*.

The neoclassical model fails us over the same landscape, however, when we confront certain health problems due to agriculture. Congenital abnormalities, breast and prostrate cancer, endocrine disruption, and non-Hodgkin's lymphoma visit these areas, and ag chemicals are highly suspect.[7] The problem is stickier, partly because epidemiological data gathering and analysis take time to ferret out how much of the health problem may be due to herbicides, fungicides, insecticides, and bactericides. These "—cides" are molecules that disrupt cells, tissues, and organs. Human health and survival are the consequence of homeostatic mechanisms evolved to surpass the threats nature has dealt us over millennia. The number of health problems derived from the human-made molecular assaults on our bodies is unknown. Now we are up against ignorance.

Decades lapsed from the time researchers became concerned about the health effects of tobacco to the era in which they began to gather data, analyze it, and make the conclusions now embedded in the surgeon general's report. Funding the army of statisticians partitioning out such confounding factors as age, sex, genotype, and lifestyle was costly. But now we do know and can partition out the health care and social costs of losing so many people to smoking-related illnesses that turn out to be many times greater than the costs of the study. A similar accounting for effects of agrichemicals will also be expensive and take longer, possibly decades, before reliable results are analyzed. Meanwhile, these chemicals will accumulate in the soil, in our waters, and in our bodies, and if the market receives a signal about the impacts in the short run, it will be faint.

Here the three-legged stool of the neoclassical model fails. Here education is the weak stick, for it depends on the "knowledge as adequate" worldview. This worldview is betting the future of the planet on the belief that our several-hundred-year-old understanding is adequate.

The Precautionary Principle

The architects of the precautionary principle have created a fine structure for dealing with most environmental problems. Unfortunately, it is inadequate here for it does not account for what sometimes happens when the scale is expanded. With increased scale, emergent properties of an unforeseen if not unforeseeable nature can appear, often when it is too late to correct the problem or even to clean up the mess. Plant breeders know that resistance to an insect or pathogen for one of their "releases" may have met the test standards in the research plots, but crash with an expansion of the variety over time and space.

Innovations often move through the testing phases conducted in research plots or on lab benches at what many consider to be a snail's pace. Once these innovations reach the commercial market, however, they can be disseminated worldwide, nearly overnight, by the global corporations controlling their distribution. For example, in little more than five growing seasons transgenic crops are grown on over 50 million hectares worldwide.[8] Although deemed safe and manageable in field plots, transgenic maize DNA has already introgressed into traditional maize land races in remote areas of Mexico.[9] The rapid expansion in scale from plot to planet requires detailed accounting guidelines based on the extent of ignorance in our hundred-year-old understanding of evolutionary-ecological systems regardless of results from even hundreds of test plots. Acknowledgment of the extent of our ignorance about the impacts of such widespread distribution in so short a period would have prevented such a rapid distribution in the first place.

With this in mind, should not our mission be to help create a new kind of accountant to help fill the gap between the bookkeepers of business and the global accountants?

Preparation for Accounting in the Little Explored Middle Ground

There are two arbitrary decisions to be made when considering a boundary. First, there is the decision as to where the boundary shall be. This is arbitrary because boundaries can be drawn at will by an observer to divide any two sets of things or places. Second, there is a decision to be made about the significance of difference found between the two sides of the boundary once it is investigated. The first decision is in abstract not involving observation; the second is a judgment relating to what is observed. The second decision assigns the boundary in question to one of two classes: natural boundaries (which seem real) or unnatural boundaries (which exist only through *a priori* definition). If the differences found between one side of the boundary and the other are judged sufficient, then the boundary could be called natural. Sometimes comparison between several alternative boundaries is needed so that the one showing the greatest differences between sides may be chosen as meeting the criterion for appearing to be real.[10]

Before the business accountants can set up the books and create the necessary columns and rows for entering the ciphers, they need to know the boundaries of consideration. These may be derived by law or public policy or some commonsense understanding handed down through tradition.

The middle-ground accountant in contrast will have to create the boundaries. Such an exercise will necessarily draw more and more of nature's designs into the realm of human purpose.

When humans (rather than God) are the creators of purpose, utility is automatic, and economics threatens to appropriate more materials, energy, and time for its kingdom. Even with our intention to "do good" by the environment, our language interacts with our thoughts and we begin to invent phrases such as "nature's services" as used earlier. Usually buttressed by equations, eventually a language develops and forces us to feel "informed" and confident of our conclusion about the state of the world. When we turn to face the world with this language (and attention equations) it often fails us. The problem will become more acute if our hardheaded utilitarianism causes us to be too dismissive of what we perceive to be the non-usable biotic and abiotic inventory of the ecosphere. The culture that bred our "knowledge-as-adequate" worldview did not teach us that before we can be very precise we have to be highly uncertain.[11]

An ignorance-based worldview is healthier and need not stop our progress toward creating the middle-ground boundaries. As we go about our business of establishing boundaries for the inventory in an ignorance-based worldview, the evolutionary-ecological perspective is our friend.[12] For example, when making a decision about whether a certain chemical molecule should be introduced, we can go beyond questions about its persistence in the environment and its efficacy and ask what evolutionary experience do our tissues have with this molecule? If the answer is none, then we would say that it should be considered guilty until proven innocent. We would not say that it will be forever regarded as guilty. Our immune system is complex and may, because of past evolutionary experience with other foreign substances, have the necessary molecular equipment to take care of the new assault. From empirical experience, however, we know that what may be on the line is a wide range of ill effects, including disruption or damage of mechanisms responsible for regulating cell division resulting in cancer.

This can all be avoided if we begin with how nature's ecosystems and species comprising them have operated on this planet over millions of years. Life forms interacting with the nonliving world and with one another using the sun's energy have been forced to live within the constraints imposed by the laws of thermodynamics.[13] The individual life forms around us today, the ecosystems of which they are a part, and the ecosphere itself, which embraces all of the earth's systems, are modern beneficiaries of the long stream of energy wars dating back to an early Earth, perhaps even to the molecular building blocks of life before there was life. A mixture of competition and cooperation has led to optimized relationships.

The Egg as an Example

We have chosen the chicken egg as the starting point for our preparatory exercise. We begin with an egg produced by a wild jungle fowl unhampered in her activities by humans (Figure 11.1). She runs on contemporary sunlight, and the material sources of her basic needs are recycled. Recycling periods for her waste products and her body itself can be measured in years, but most often in less time. Her complex immune system, evolved over millennia to fit her complex surroundings, provides the antibodies against disease. Once mature, she will produce some fifty eggs per year. The predators that steal her eggs or eat her chicks balance her fecundity.

Humans as part of the hen's ecosystem can "steal" some of her eggs and have a similar effect as a snake that would do the same. If humans collect too many eggs, they could push the wild fowl toward extinction. If humans collect too few, they could be left wanting. The problem posed by introducing egg needs into the picture is well stated by Wendell Berry in *Home Economics*:

> Humans, like all other creatures, must make a difference; otherwise they cannot live. But unlike other creatures, humans must make a choice as to the kind and scale of the difference they make. If they choose to make too small a difference, they diminish their humanity. If they choose to make too great a difference, they diminish nature and narrow their subsequent choices; ultimately they diminish or destroy themselves.[14]

Through domestication we have bred and organized our chickens to provide a more stable egg supply for our growing population over the long term despite its more extractive nature as compared with wild egg production (Figure 11.2). Domestic egg production took place in the back yards of people the world over during the past 8,000 to 10,000 years. Reliable production must have become the desired norm early on. Even within modern industrial societies, backyard domestic egg production relies heavily on the hen's ability to fend for herself during the day eating insects, gleaning spilled grain, and recycling her keeper's food scraps. A simple structure built of wood and metal can provide protection from predators at night. This type of production has replaced little of the hen's natural ability, yet has increased production from a yearly average of approximately one per week in the wild to around three per week.

Figure 11.1. The wild egg.

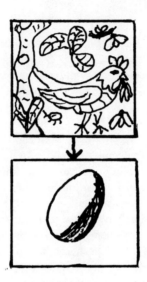

Figure 11.2. The domestic egg.

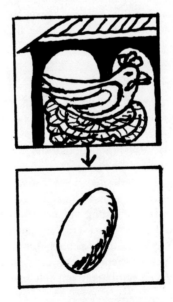

Even though the negative environmental effects of backyard, domestic egg production must have been there from the beginning of domestication, they are nothing close to the scale of the modern industrial egg production system (Figure 11.3). Over the last hundred years, the scale of organization and effects of egg production have increased dramatically through specialization affordable only with fossil fuel. The industrial mind has promoted a narrowing of both purpose and function in the chicken and egg enterprise. Chickens intended for meat production are not used for egg laying. Hens intended for egg laying have that primary purpose. Neither are available to manage barnyard pests or to recycle the farmhouse food scraps.

Since our industrial layer is not intended for multiple uses, the means of production can be organized in smaller and smaller spaces. Confinement allows for a complicated supply and distribution system to be more time efficient. The foraging and defense capabilities of our wild fowl are now unnecessary. Eggs are directly laid on conveyor belts. Beaks must be trimmed to prevent the cannibalism that confinement fosters. Much of the hen's immune system, formerly operating under free-range conditions, is now obsolete. The industrial content makes our chicken more susceptible to disease, requiring frequent prophylactic administration of antibiotics even though the wild-sponsored immune system may still be intact in open foraging. The benefit: four eggs per week per chicken rather than three or one.

These three scenarios—the wild egg, the domestic egg, and the industrial egg—present examples of the kind of reality our student of boundaries will face in trying to answer the simple question: What is the kind and scale of difference we shall make? How do we maintain our humanity *and* the integrity of the planet? Answers to these questions will combine narrative with number-smithing for such considerations as caloric production, greenhouse gas emissions from production practices, and fuel use for grain production. These calculations will then be used to compare one system with the other and enable decisions: Do we buy an industrial egg, raise our own in the backyard, or gather from the wild? Before estimates can begin, before number-smithing becomes the order of the day, we will be defining boundaries.

Figures 11.1–11.3 are flow diagrams for our three eggs and represent the type used to consider energy, nutrient, and material budgets. Figures 11.4–11.8 represent subsets of specific activities necessary to carry out those in other figures. References to Figures 11.4–11.8 are found in the boxes shown in Figures 11.1–11.3 or are indicated by corresponding symbols. These simplified representations of complex systems will help illustrate some of the boundary problems our accountant must deal with. Also, we believe that discussion of even these simple diagrams clearly indicates the need to develop a taxonomy of boundaries.

Figure 11.3. The industrial egg.

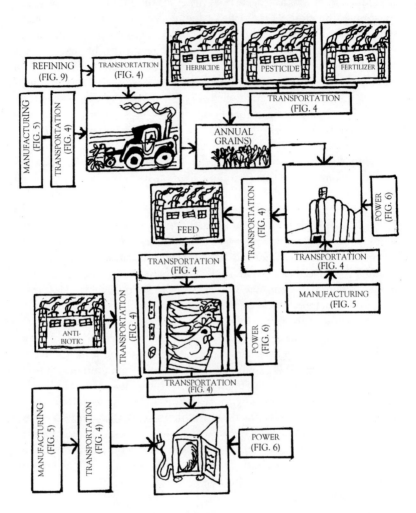

We will begin with Figure 11.3 because it contains more representations and is more complicated than Figures 11.1 and 11.2. An initial consideration is that the map is not the reality, only the assist to both boundary selection and accounting. If the accountant knows the map well and the territory poorly, the age-old problem of the abstraction overriding the particular will cause the accountant's veracity as a middle-ground accountant to collapse.

A second boundary consideration: The accountant will notice that there are subsets within subsets within subsets *ad infinitum*. Figure 11.4, for example, illustrates the reliance of the transportation industry on itself in transporting the necessary parts and fuels to where they are used. Where in this regression do we stop? In considering the manufacturing process producing the refrigerator in which the industrial egg is stored, shall the accountant put on the books the factory that produced the parts to make the factory that produced the refrigerator? The refrigerator, at least, must be considered, because the industrial egg under consideration has likely traveled nearly one

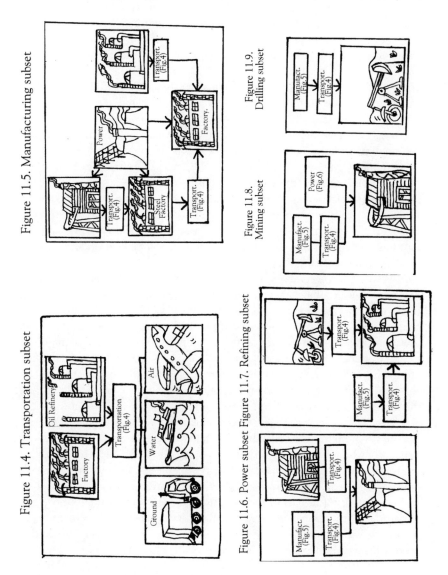

Figure 11.5. Manufacturing subset

Figure 11.9. Drilling subset

Figure 11.8. Mining subset

Figure 11.4. Transportation subset

Figure 11.6. Power subset   Figure 11.7. Refining subset

thousand miles from the factory in which it was laid. Eggs don't keep well for the long periods of time required in this transportation and marketing system without refrigeration. Similar considerations must be made to determine where the boundaries will be declared to account for the manufacture of drilling rigs and tractors standing behind the egg.

How far back in time should our consideration run? We can be ridiculous and go back to the Paleozoic, some 225 million years ago. With this time frame, the current extraction and burning of fossil fuels is a form of recycling without our boundary of consideration and so could be regarded as positive activities. This gets us nowhere and so we come to what we call human time, which is embedded within and dependent on the products of geologic time. Though we have had the big brain 150,000–200,000 years and agriculture 10,000 years, even ten millennia seems too long, given that the human experiment, as it is currently being run, is better measured in decades than in centuries. In the case of fossil fuel consumption, decades seem too short, but centuries or a few millennia about right given that the residence time of a carbon dioxide molecule in the atmosphere is 50 to 100 years.

Jumping from Figure 11.1 to Figures 11.2 and 11.3, it is evident that the number of boundaries that must be considered by the accountant greatly increases. If the boundaries represented in the schematic are where important effects occur, then there will be more for the accountant to enter on the books for the industrial egg than for the domestic or wild egg. The accountant has the burden to choose which boundaries will be kept under consideration for the eventual tally.

We are concerned about what is not shown on the schematics. There are questions that may have answers that will change our boundary. For example, in Figure 11.3, assume the egg is the last of its kind. What is the condition of the people who will collect the egg? If this is the last or one of the last eggs, then our accountant must stretch the temporal boundary of consideration to ensure that there will be eggs for tomorrow or at least that there will be a hen around to produce the eggs for the day after tomorrow. If the people relying on this method of egg production are hungry, in poor health, or both, then their diminished state must be taken into account and the boundary adjusted.

To continue this thought experiment, assume the marginal difference between the diminished and fulfilled human state is an extra three eggs per hen each week. If avoiding a diminished state is important and a fulfilled existence relies on a hen producing four eggs per week, then the accountant's boundaries of consideration would be those arcing out from the industrial egg—assuming there is no other way to get the required number of eggs. What is not represented by the diagram is now compounded, perhaps exponentially. *The diagram of the wild egg masks important boundaries, but the diagram of the industrial egg masks far more.*

To illustrate the problem of masked boundaries and how the problem dramatically worsens in complicated systems, consider the industrial hen's food production at its source—the field. Under modern production practices, soil and nutrient loss, fossil fuel dependency, generation of greenhouse gases, poisoning of wildlife, and reduction of biodiversity must be taken into account—boundaries must be drawn. This will be a difficult if not impossible challenge to meet as each of these processes produces effects at different boundaries of space and time.

Complicating it more, the effects of these processes must be considered in the context of the entire system in which they are involved. For example, typical soil and nutrient losses from one field in the Midwest in one year may have no calculable effect that could be put on the accountant's books. The particular soil and nutrient losses from that field in that year together with the soil and nutrient losses from all the fields draining into the same watershed does, however, result in downstream hypoxic zones such as the dead zone in the Gulf of Mexico. The management of one particular field is a consequence of the dominant way of farming in the Midwest. The soil and nutrient loss in one year together with the losses before and after are irreversible. Effects of managing that field in one year must be considered in the context of several years.

The accountant, caught up in these intricate field-level boundary considerations, may forget the consideration given to the diminished humanity of our wild egg collectors discussed earlier. Beyond the field, what is the condition of the farm family producing the grain to be transported to the pellet mill to be processed and transported to the chicken factory? People along the way are currently off the books and masked behind the complicated schematics of the simplifying diagram. Through either neglect or the temptation to simplify, the opportunity to drop important effects from the books increases as the number of processes that humans must manipulate and consider increases.

Preemptive accounting might warn us off the industrial egg production track due to the likelihood that important effects will be left out of the bookkeeping process. We might then look to domestic production for our egg needs. Although our domestic hen produces only three eggs each week, a full production accounting might indicate that the backyard, "multitasking" hen reaps benefits for the owner not counted solely in terms of eggs. Writing in the 1930s, Dr. A. L. Hagedoorn researched and advised farmers and animal breeders to be mindful of what he called the "symbiosis-group" or "symbiotic unit" within which the creature of interest is found. In his text *Animal Breeding*, first published in 1939, he said:

> At first sight a breed of poultry, in which the hens lay 200 eggs yearly, is vastly superior to a breed in which the hens lay 50, and yet the cheapest

eggs are probably produced by such hens as the native poultry in China and Java, where everybody keeps only a few hens, and nobody ever thinks of giving them any food at all.[15]

Gene Logsdon offers a more savory firsthand account of the symbiotic unit described by Dr. Hagedoorn. His description of the relationship between his cows and his chickens, among his cows, and the barnyard insects is an example of a coherent component of the farm agroecosystem:

> During the summer, my chickens get most of their protein supplement by eating blood-gorged flies off the cows. When I let the hens out in the morning, they scurry off to where the two cows lie in the shade, first attacking the big, slow horseflies they can reach from the ground, sometimes jumping up on the reclining cows in their eagerness. When the horsefly population is depleted momentarily, the hens linger around the cow's heads, nabbing the nimbler face flies. The cows never budge during any of these maneuvers, obviously aware not only of the benefits accruing to them but also trusting completely a chicken's ability to peck within an eyelash of their eyeballs without injuring them.
>
> When the chickens tired of chasing flies, they attacked the cows' manure—what we call cowpies—a normal occupation by chickens and hogs, but one which I never before took the time to observe closely. The hens tore apart the cowpies, reducing their bulk by a third, scattering the fecal matter over a broader area so that the grass was not smothered under it. From the cowpies, seemingly as delectable to chickens as apple pie is to humans, the hens ate partially digested grains and they greedily gobbled chunks of digested grass. The hens were not starved to this diet. They had plenty of fresh grain and grass available to them. They preferred this food. In addition, they pecked at tiny specks of indiscernible stuff and fly eggs, plus other insects drawn to the manure, not to mention earthworms that had worked up from below to dine on the succulent organic matter. Over the course of a summer, I counted seventeen distinct species of insects in and about the pasture cowpies, all working out their life cycles in some kind of tenuous connection with the manure.[16]

Some might feel this insightful narrative with its rich consideration of farmyard boundaries decides the case in favor of the domestic egg. We certainly feel it's more right than an industrial egg. We urge, however, a broader assessment of boundaries beyond the farmyard to include the resilience of the system over time and, once more, the condition of the people collecting the eggs. Nervousness about being wrong will better serve us over the long term than comfort in feeling we're right.

This exercise leaves many questions unanswered and vague notions of the boundaries for appropriate bookkeeping. We have tried no formal language to communicate the relatedness of one boundary to another. Plant and animal taxonomy has served the biological disciplines in a formal way. It evolved and continues to evolve over time. Without an established boundary taxonomy we are left without rigor in assessing the options we have before us, whether they involve eggs or energy, and are left susceptible to a power structure that never gets asked, "Where are the numbers?"

Examples of the Insufficiency of Current Boundary Language

Two current issues illustrate the need for a thorough consideration of boundaries rather than simply relying on what feels good. Since the Kyoto Protocol, plans have been called for to sequester more carbon in agricultural soils to mitigate the global warming potential of carbon dioxide emissions. No-till agricultural production of annual crops has been proposed as a viable option and has indeed been shown to sequester some soil carbon.[17] If we study only the farm, we might conclude that there is a net decrease in atmospheric carbon and thus a mitigation of global warming potential. But of course the boundary of the farm is too narrow. Carbon and nitrogen emissions to the atmosphere, resulting from the production, transportation, and application of the fertilizer, more than negate the mitigating effects of the sequestered carbon. No-till production of annual crops, therefore, is a net contributor to global warming. As William Schlesinger, an "accountant" in this case, points out: "Treaty negotiators must keep in mind the complexities of a full accounting of the carbon emissions and sinks associated with various human activities."[18]

The push for ethanol production from crops is another example. The federal government is providing tax incentives to promote the production of ethanol from crops for transportation. In the late 1970s, the Land Institute conducted a study to consider the consequences of a massive alcohol fuels program to meet on-farm energy demands, about one percent of the U.S. total at that time. We assumed a solar still that would operate at top theoretical efficiency. The spent mash that is left over is high in protein so we assumed that this potential livestock feed could be given an energy credit equal to the amount of energy to grow an equivalent amount of protein in the soybean crop. The year we did the study there were 70 million acres of corn producing, on average, 100 bushels per acre.

The result: in order to meet on-farm traction energy needs, which amounted to one percent of the total oil used in the United States, not 70 million acres of 100 bushels per acre corn but 117 million acres were needed. And that is leaving nothing for livestock or people. In addition, the rate of soil erosion would increase by more than thirty percent since the extra forty-seven million acres of production would be on poor quality soils and steep landscape positions. If instead of using crops for ethanol production, were we to increase the load factor in the American automobile from 2.2 to 2.4 people per passenger mile, we would have an energy savings equal to 117 million acres of corn. In other words, a massive alcohol fuels program would not do what a modest conservation program could accomplish.

We mention these two examples relating to government policy to illustrate the necessity of being specific in our discussion of boundaries. If we restrict our accounting to the field level, no-till management may well sequester more carbon than the conventional system. The preemptive accountant, however, must not only extend the boundaries of consideration outward, but must also be able to communicate to others specifically which boundaries will be considered. This type of accounting in the "middle ground" is currently being ignored in large part due to the lack of appropriate boundary determinations and the lack of a defined language about boundaries. Boundary taxonomy could provide the necessary supporting language.

Concluding and Summary Remarks

Dan Lutten, professor emeritus of geography at the University of California, Berkeley, once described a system as "an object of interest together with its significant environment."[19] The "significant environment" statement implies a boundary, probably an artificial boundary. If our object of interest is a candle, then our "system" requires a flame source, a flame, a wick, wax or the equivalent, a place to position the candle, and gravity. The mental exercise forces us to consider variables we are likely to ignore during a candlelight dinner. Even when the exercise is to define the system, most will not have considered the necessity of gravity.

So here is the point. Students of boundaries will become discoverers of significant variables that we all take for granted. They will be discoverers and communicators of where a "significant environment" begins and ends in terms of our goal of reducing the impacts we humans have on the ecosphere. Furthermore, it will be their task to find the big contributors in the system. Because our students of boundaries will not be working alone and will eventually need to communicate their results to the broader public, they will need a systematic language in which to articulate their understanding. A taxonomic system, such as those developed for other disciplines, could provide the linguistic framework to keep our boundary accountants exercising a common language among their peers and the public.

Boundary creations, accounting, and number smithing can serve to raise the level of awareness as to what counts and how much. We consider the proposed exercise to be essential if we are to nest the human economy within nature's economy.

Notes

1. Plato and Aristotle lamented the land degradation problem of Greece. The paleo-Indians left no templates to record their perception of loss. To account for a problem through some form of expression, either with a cipher or through narrative, is the necessary first step for either halting or correcting the offense or both.

2. P. M. Vitousek, P. R. Ehrlich, A. H. Ehrlich, and P. A. Matson, "Human Appropriation of the Products of Photosynthesis," *Bioscience* 36 (1986): 368–373.

3. R. Lal, R. F. Follett, J. Kimble, and C. V. Cole, "Managing U.S. Cropland to Sequester Carbon in Soil," *J. Soil and Water Conservation* (1st quarter, 1999): 374–381.

4. Nancy Rabalais and R. E. Turner, "Hypoxia in the Northern Gulf of Mexico: Description Causes and Change," in *2001: Coastal Hypoxia: Consequences for Living Resources and Ecosystems*, Coastal and Estuarine Studies 58. Washington, DC: American Geophysical Union, 2001, 1–36.

5. R. Constanza, R. d'Arge, R. de Groot, S. Farber, M. Grasso, B. Hannon, K. Limburg, S. Naeem, R. O'Neill, J. Paruelo, R. Raskin, P. Sutton, and M. van den Belt, "The Value of the World's Ecosystem Services and Natural Capital," *Nature* 387 (1997): 253–259.

6. Professor Charles Washburn has asked the question, "At what point did the value of nature's services exceed the value of the output." At the beginning of agriculture, in the very early days of the development of techniques beyond the use of stone tools, the value of nature's services must have been very small compared to the value of a product combining new technique and labor. By the time of Charlemagne, around A.D. 800, the onslaught against the forests was well under way and had already nearly finished in the Middle East. It seems safe to say that some time between A.D. 1 and 1800, before the industrial revolution took off, the crossover had already occurred. This means that in the 8,000 years that had passed since agriculture began, the net product may have exceeded the value of the capital asset and deficit spending became the rule.

7. L. Hardell and M. Eriksson, "A Case-Control Study of Non-Hodgkin Lymphoma and Exposure to Pesticides," *Cancer* 85 (1999): 1353–1360. C. Lu, R. A. Fenske, N. J. Simcox, and D. Kalman, "Pesticide Exposure of Children in an Agricultural Community: Evidence of Household Proximity to Farmland and Take-Home Exposure Pathways," *Environmental Research* 84 (2000): 290–302. E. M. Bell, I. Hertz Picciotto, and J. J. Beaumont, "A Case-Control Study of Pesticides and Fetal Death Due to Congenital Abnormalities," *Epidemiology* 12 (2001): 148–156.

8. C. James, "Global Review of Commercialized Transgenic Crops: 2001," *ISAAA Briefs No. 24: Preview*. Ithaca, NY: International Society for the Acquisition of Agrobiotech Applications, 2001.

9. David Quist and Ignacio Chapela, "Transgenic DNA Introgressed into Traditional Maize Landraces in Oaxaca, Mexico," *Nature* 414 (2001): 541–543.

10. T. F. H. Allen and Thomas B. Starr, *Hierarchy: Perspectives for Ecological Complexity*. Chicago: University of Chicago Press, 1982.

11. Ask ten students consisting of five groups of two to measure the length and width of a room approximately 7 meters by 10 meters using a tape marked off in millimeters. Tell the students that we want to be certain of the measurement within a millimeter. At the end of the exercise all ten are likely to agree with the statement stated in the text.

12. It is increasingly difficult to talk about one of those fields without talking about the other. At this point in history, we lack yet need one term that captures both.

13. Neoclassical economics has yet to adequately incorporate the painful reality of the entropy law. It is a cultural problem in that the idea that nature should be subdued or ignored may go back as far as our gathering-hunting past, to the upper Paleolithic and even before. We could get away with destructive activity then. More ecosystems were intact and we were few in number. In the modern industrial world, that attitude is no longer viable. In fact, it is the attitude responsible for the modern environmental problems. At this point, the evolutionary-ecological perspective becomes available for providing a direction toward an evaluation of a different brand of economics.

14. Wendell Berry, *Home Economics*. San Francisco: North Point Press, 1987, 7.

15. A. L. Hagedoorn, *Animal Breeding*. London: Crosby Lockwood & Sons, 1939; 3rd ed. 1948, 60. See also 65.

16. Gene Logsdon, "The Importance of Traditional Farming Practices for a Sustainable Modern Agriculture," in *Meeting the Expectations of the Land: Essays in Sustainable Agriculture and Stewardship*, eds. Wes Jackson, Wendell Barry, and Bruce Colman. San Francisco: North Pont Press, 1984, 4–5.

17. G. Philip Robertson, Eldor Paul, and Richard Harwood, "Greenhouse Gasses in Intensive Agriculture: Contributions of Individual Gases to the Radiative Forcing of the Atmosphere," *Science* 289 (2000): 1922–1925.

18. W. H. Schlesinger, "Carbon Sequestration in Soils," *Science* 284 (1999): 2095.

19. Personal communication with the first author.

Chapter 12

How to do Things with Food: a Plea for Multiple Ontologies

Bruce Hirsch

Starting Points

Our modern industrial agriculture and food system creates significant threats to the environment and public health. The system is unsustainable, but there is no broad public outcry or large-scale social movement to change production and distribution practices that depopulate rural communities and separate urban residents from their sources of food. A small group of distinguished writers, scientists, and activists is working to reform the industrial agriculture system on the model of healthy ecosystems that restore and protect clean air, clean water, living soil, and rich biodiversity, while promoting wise energy use, protecting public health, and redeveloping vibrant rural areas.

While we can describe these goals, we have trouble finding a way to reach them. Our starting points seem to be wrong. The way we frame the issues, the way we include some items and omit others limits our progress. Should we start with the hypoxic "dead zone" in the Gulf of Mexico, estimated to be the size of New Jersey? What about the loss of the Ogallala aquifer as a source of irrigation for the central United States or the destruction of topsoil, a nonrenewable resource more precious than fossil fuels? Can we ignore the contribution that industrial agriculture makes to insect resistance to pesticides or the prevalence of "diseases of affluence" such as cardiovascular illness?[1]

As a society, we are just beginning to frame the issues of agriculture and the food system in a way that can include these problems. This framing activity defines the boundaries of consideration, and these boundaries will determine which problems we identify, how we describe them, and what, if anything, we do about them.[2] We have an opportunity to limit or expand the way we think about moving toward a more sustainable agriculture and food system.

I want to rethink these boundaries of consideration in several ways. I will try to throw light on the role of public interest organizations in framing the basic structures of our agricultural system and how this role can contribute to change most effectively. Specifically, I consider the role of civil society institutions such as philanthropic foundations in initiating change. These foundations and the public interest organizations they support have done a significant amount of work to understand the challenges in reforming the industrial agriculture and food system. Their role in this field is still developing, but they can move more quickly than the government to identify problems and propose solutions, and they can act in the public interest without fear of being voted out of office or failing to promote the interests of shareholders. These institutions have an opportunity to function as "society's passing gear" to accelerate progress in solving systemic problems.[3]

As part of this effort to rethink the boundaries of consideration for change in agriculture and food systems, this chapter also examines the early work of Martin Heidegger, particularly his analysis of human existence, culture, and nature. I look to Heidegger for the same reason writers and activists consult the work of Leopold or Thoreau, to find a fundamental relationship between humans and nature. As we shall see, Heidegger's contribution is challenging, but worthwhile. Writing in the early twentieth-century, he described how our everyday experience is mediated through culture and technology in ways that have the potential to separate us from nature and devalue it.

But I do not claim that Heidegger was an environmentalist in any contemporary understanding of the term. And the historical record clearly shows that he was not a democrat. The value of his early work lies in extending his original analysis using our twenty-first-century experience of phenomena similar enough to those he described. Heidegger shows how our various ways of being involved with the world create their own boundaries of consideration and how these boundaries generate further opportunities for human engagement in the diverse worlds that we inhabit.

Heidegger offers a description of the world that challenges the dominant Cartesian account of independent subjects and objects, replacing it with a network of cultural significance and social practice that allows objects to be what they are within a variety of contexts. Across these contexts, the culture will value some practices and meanings more than others, generating a body of received wisdom that limits our ability to experience the richness and diversity of the world. In everyday living these calcified cultural pathways are not objects of reflection or cognition. Modern life would be impossible if we stopped to analyze them regularly, but they continue to have a pervasive, amniotic character for us.[4]

Sometimes our amniotic world breaks down and we have to step outside of it to recognize and repair it. This gives birth to the Cartesian, scientific approach to the world, which is both derivative and necessary in the West. If we are to recognize and address our environmental problems as a culture, we'll need our best interdisciplinary scientists, organizational and communications consultants, policy-makers, artists, and writers to help us establish a new network of institutions, policies, and practices. The public interest community has been working on these issues and bears a special responsibility for identifying the boundaries where change can take place.

Entry Points

Over the last few years the public interest community has shown increasing concern about the industrial agriculture and food system. Foundations that do not see themselves as "agriculture funders" and organizations not exclusively dedicated to agriculture are involved in issues relevant to farming and food systems. These are issues in which the industrial system is reaching its limits, forming boundaries that show where it is breaking down, despite or because of its overwhelming economic dominance.

Emerging Boundaries of the Industrial System Where We Find Public Interest Activity

- The threats to water, air, soil, and human health from large animal confinement operations in several states
- The risks to ecosystem and human health from the nontherapeutic use of antibiotics, particularly those that are important to human medicine
- The health risks associated with pesticide use, including cancers, compromised reproductive and neurological systems, and developmental problems in children
- The contribution that industrial farming makes to global climate change
- The significant ecological footprint created by industrial agriculture's use of resources, including water, soil, and fossil fuel energy
- The efforts of some corporations to market unsustainable applications of biotechnology as an answer to the problems created by industrial agriculture

- The need for a public interest research and policy agenda that outlines steps toward achieving an ecologically sound, socially just, and economically viable food system
- The redesign of university extension programs to help farmers make the transition to organic and sustainable systems
- The development of regional food systems that participate in local and global markets while adhering to high environmental and community standards
- Recognition for the value of and public interest in protecting and renewing common assets that provide essential ecosystem services
- The reform of international trade policies that promote cheap globalized food that hides the true costs of industrial production and destroys the ability of farmers in developing countries to feed themselves and their neighbors

Now although the entry point for some of this work is an interest in agriculture and food systems, most funders come to these topics from other points of view.

Entry Points for Emerging Boundaries

- The polluter pays principle
- The precautionary principle
- The right-to-know-principle
- The promotion of science in the public interest
- The development and adoption of alternative methods of risk assessment
- The implementation and enforcement of clean air and water regulations
- The promotion of environmental and social justice

Widespread adoption of environmentally conscious agricultural production and distribution systems has the potential to address big environmental and public health challenges in settings where grantmakers seek to influence policy and practice. For some time, European practitioners have spoken of the "multifunctionality" of ecologically sound agriculture and food systems.[5] By this they mean that since modern industrial agriculture has generated significant environmental threats, its reform based on a model of healthy ecosystems has the potential to restore and protect the fundamental conditions for nonhuman and human life. A short list of goals to which sustainable agriculture could contribute includes the following:

Areas Where Funders Seek to Influence Policy and Practice

- The restoration and maintenance of biodiversity
- The provision of healthy soil and clean air and water
- The preservation of wilderness
- The careful use of energy and the development of renewable energy
- The protection of public health, including children's and farm workers' health
- The economic and social viability of rural communities
- Food security for vulnerable populations
- The protection of regions against sprawl
- The building of new economic relationships between urban and rural residents

We want an agriculture and food system that contributes to *Policy and Practice* goals, but can we achieve it with the kind of starting points in *Emerging Boundaries* and *Entry Points*? The voices of some of our most decorated grassroots activists and writers suggest that changes in policy and practice don't follow directly from philanthropic guidelines, statements of principle, or environmental ethics. David Abram suggests that the reclamation of the full-blooded holistic experience of the natural world can restore our respect for human and nonhuman life. He recommends "a renewed attentiveness to [the] perceptual dimension that underlies all our logics, through a rejuvenation of our carnal, sensorial empathy with the living land that sustains us."[6] The boundaries of human experience do influence what counts as important for individuals and societies. What kinds of boundaries are significant for progress in changing the current industrialized food system?

In the following section I attempt to answer this question by sketching the boundaries of human existence, culture, and nature that follow from the early work of Martin Heidegger. These boundaries account for certain kinds of human engagement in the world, engagements that describe who we are and what we do within cultures and subcultures. If I am right about the value of looking to Heidegger for understanding boundaries and the ways human beings involve themselves in the world, the result may be some insight into how change occurs, in food systems and other areas.

Boundaries, Diversity, and Transparency

In seeking to understand Heidegger's comments on nature, I focus on his masterwork, *Being and Time*.[7] Heidegger launches an attack on the epistemological foundations of Western philosophy and science with his description of human existence as being-in-the-world. Being-in-the-world is meant to replace the familiar subject–object, mind–world model that separates human subjectivity from an independent world with which it must somehow make contact.[8] On Heidegger's view, human experience is from the start embedded, saturated, and engaged with the world. The world is infused with human-made significance and humans are fused to this world: there is no bare subject confronting isolated objects. Heidegger's description of being-in-the-world offers a holistic a priori background for our familiar, second-order Cartesian-colored subject–object distinctions.

Heidegger's move has an intuitive appeal for environmentalists.[9] If this precognitive boundary includes the natural world, perhaps our society can recover an ethical imperative that supports environmental protection and restoration. Such an ethical imperative, if harnessed to an organized social movement, would no doubt encourage a food system that treats plants and animals, including humans, differently than industrialized agriculture does. We would then be reevaluating our usual boundaries of consideration and staking out new ones. The *Emerging Boundaries* would take on greater significance; their *Entry Points* would gain greater public support.

But Heidegger's picture is more complicated. When he jettisons the static, disinterested subject, he opens up a dynamic field of engagement for human activity. In *Being and Time*, we encounter nature through different kinds of engagement with the world.

- As farmers we see natural events as signs for other natural events of importance to us: the south wind is a sign of rain.[10]
- When we devise methods for measuring time, we take into account the Earth's movement around the sun and the changes between day and night.[11]
- In writing and studying history we recognize the role nature plays. Heidegger says, "nature is historical as a countryside, as an area that has been colonized or exploited, as a battlefield."[12]
- As investigators into nature, we encounter it through the natural and mathematical sciences, the Cartesian ideal whose supremacy Heidegger is interested in undermining.[13]
- As users of tools and equipment, we discover their origins in nature through raw materials such as wood, iron, water, wind, and rock.[14]
- As users of tools and equipment, we see nature as an obstacle to their employment for our ends. "A covered railway platform takes account of bad weather," allowing us to achieve our goal of comfortable travel.[15]

- As artists and poets, we confront the majesty and beauty of nature. Heidegger refers to "the Nature that 'stirs and strives,' which assails us as landscape." He adds that "the botanist's plants are not the flowers of the hedgerow; the 'source' which the geographer establishes for a river is not the 'springhead in the dale.'"[16]

Heidegger's point is that the kinds of engagements we choose determine how natural objects in the world and the world itself is disclosed, encountered, or revealed. But the problem is that we are likely to take the dominance of some disclosing involvements as defining the world in an exclusive and correct way. Heidegger calls this the "leveling down" of the available interpretations for human engagement in the world.[17] In our culture, this happens when we tend to engage in practices that disclose the natural world as primarily a bundle of raw materials for industrial processes, including agriculture. When we narrow the boundaries of consideration in this way, we marginalize the *Emerging Boundaries* and *Entry Points*, blocking access to *Policy and Practice* goals.

As societies adopt a dominant approach to public engagement, they build a corresponding complex of social, cultural, and practical arrangements to support it. This complex is transparent to the public as long as things proceed smoothly. Heidegger starts with his famous example of the hammering activity taking place in a workshop.[18] When hammering proceeds without interruption, the background that makes it possible is not our focus; it is there but transparent. But when the hammer breaks or is unavailable, the network of practical relations in the workshop becomes the focus of repair. Thus, in *Being and Time*, the world is disclosed to us through various equipmental complexes. These equipmental complexes constitute the various boundaries of consideration and corresponding social and cultural arrangements by which we interpret, analyze, and seek solutions to environmental problems. The insight can be extended in a significant way.

The Economist recently published "A Survey of Information Technology" that considers new developments in the field.[19] The survey includes an article with the decidedly Heideggerian title, "Now you see it, now you don't: To be truly successful, a complex technology needs to 'disappear.'"[20] The article briefly traces the evolution of the automobile from an engineering curiosity to an indispensable feature of modern life made possible by a simplified user interface that requires no more engineering insight than the ability to turn the key. This "migration of complexity" is a regular feature of technology, according to Marc Benioff of Salesforce.com, and

> echoes the process of civilisation. Thus, every house initially has its own well and later its own generator. Civilisation turns houses into 'nodes' on a

public network that householders draw on. But the 'interface'—the water tap, the toilet flush, the power switch—has to be 'incredibly simple.' All the management of complexity now takes place within the network, so that consumers no longer even know when their electricity or water company upgrades its technology. Thus, from the user's point of view...technology goes through a gradual disappearance process.[21]

In the twenty-first-century there are obvious examples where a transparent interface created by an appropriate migration of complexity is advantageous. For example, police officers in San Jose, California, complained about the difficulties in using their new car-based computer dispatch system, especially when they need to call for backup assistance. As reported by the *New York Times*, officers expressed concern about "the user interface, the all-important gateway between person and machine...a dizzying array of buttons or keys that have to be used in combinations."[22] When police need to make emergency calls, we want a transparent interface. But when it comes to the relationships between human beings and the industrial food system, smooth interfaces amplify environmental and public health threats. Our laissez-faire attitude toward the economic, political, and social distance between eaters and farmers, the shrinking share of the food dollar that stays on the farm, the absence of labeling for genetically modified foods, and, above all, the fact that urban residents can cruise through the supermarket without recognizing any of this, demonstrates the dangers of transparent interfaces. In short, when it comes to the food system, smooth interfaces retard the development of a broader social movement for change.

From this perspective, advocates for sustainable agriculture are fighting to make the industrial system's *Emerging Boundaries* (re)appear so they can enter the public dialogue and eventually be addressed. As supporters of Slow Food recognize, "It was no longer enough to know good food: now we needed to know where it came from, who produced it, and how we can ensure a secure future for its existence."[23] and "we should develop direct patterns of distribution, abolish export subsidies, and use that money to support regional markets."[24] Knowing where your food comes from (i.e., knowing your farmer) is a first, simple step in halting the migration of complexity encouraged by the industrial system. But as the Slow Food folks recognize, it's not enough; new networks of culturally sanctioned practice and significance must be built.

As in any movement for change, these new networks have innovators, engineers, and entrepreneurs who develop alternatives to the existing networks; early adopters who participate as consumers, and a range of organizations that shape a new agricultural world. This work can be encouraged by the public interest sector with the participation of grantmaking foundations that: (a) recognize the importance of developing critical attitudes toward technologies that separate us from nature, (b) communicate *Emerging Boundaries* and *Policy and Practice* goals effectively to the public, and (c) develop policies and infrastructure for the new food system networks.

A Plea for Multiple Ontologies

In her case study of a cancer cluster in Point Hope, Alaska, in the 1980s and 1990s, Nelta Edwards describes the efforts of the Inupiat community to convince authorities that they are suffering from a new health threat unrelated to the need for better cancer screening and the end to tobacco use.[25] As they try to find an explanation for the increased rate of human cancer diagnoses and health problems in the local wildlife population, health workers and the local community are unaware of government radiation experiments conducted nearby in the 1950s. Eventually, a researcher learns of the radiation experiments, but public health officials believe the levels of radiation used in the experiments were too low to change the cancer rate and conclude that the amount of radiation released into the environment was "safe."

Edwards considers three ways to resolve the impasse between the community and health officials. The first would be to do better science by collecting more data in the hope that the causes of the illnesses could be identified. This approach assumes that additional data will identify problems with the methodology used in previous studies, which found no new causes for local health problems. It doesn't alter the original boundaries of consideration; it admits the possibility of poor implementation within them. The second approach is to acknowledge that scientific language and methods as well as political considerations can distort the results of an investigation. Public health authorities may be more likely to focus on previously identified risks, such as personal tobacco use, rather than raise questions about environmental carcinogens. In addition, although health researchers found the local cancer rate to be thirty-eight percent higher than the U.S. standard rate, the small population of the region makes the difference statistically insignificant. This second approach allows that an investigation takes place within cultural contexts that affect its boundaries of consideration and its outcome, but it fails to expand the boundaries. Edwards concludes that trying to do better science or acknowledging the distortions in science both "rely on a particular ontology or understanding of reality. They rely on a realist ontology that assumes a reality to be discovered 'out there,' even though this reality may be hidden or distorted. This is also the ontology of science. Scientists set out to 'discover' an already-existing, or an objective, reality."[26]

Edwards considers a third possibility for bridging the impasse in this case. This is to take a social constructivist position, recognizing that scientific language is the construction rather than the representation of reality. She claims that social constructivism is based on an "antirealist ontology," one that denies a reality "out there" waiting to be discovered; instead we acknowledge a world of "distortion and flux." For example, official talk about "safe" levels of radiation masks the disputes within the scientific community about the effects of low doses on cells as well as the value judgment that some level of exposure is "permissible" to gain the diagnostic benefits of Xrays. In order to avoid the obvious objection that the antirealism of social construction denies physical realities such as radiation and illness, Edwards suggests an appeal to "mobile ontologies," a position that allows us to take "an antirealist stance toward 'facts' while concurrently taking the contradictory realist stance that 'facts,' or an objective reality does exist."[27]

I agree with Edwards's claim that we need a new ontology "because it may be the realist ontology of science that portends ecological destruction."[28] But the factual neodualism she proposes fails to capture the richness of nature and our experience of it, disclosed through active engagement in science or traditional Inupiat subsistence culture. In Heideggerian terms, the initial impasse was generated because the Inupiat engagement with nature disclosed its spiritual, symbiotic, and sustaining features, those not captured by Cartesian or factual dualisms. Heidegger's solution, a description of the *multiple ontologies* of nature and the world, widens our boundaries of consideration by accounting for our involvement in nature, its physical characteristics, and the cultural values embedded in our experience. Ontological dualisms miss something indispensable for environmental progress—whether it is changing the agriculture and food system or recognizing the valuable knowledge of traditional societies. Dualistic ontologies forget the role of the human actor and, since the point of environmental progress is to recognize and redress the damaging consequences of human action, they are a poor basis for epistemology and learning. Organizational consultant Peter Senge recognizes the practical import of this Heideggerian insight when he says

> From the systems perspective, the human actor is part of the feedback process, not standing apart from it. This represents a profound shift in awareness. It allows us to see how we are continually both influenced by and influencing our reality. It is the shift in awareness so ardently advocated by ecologists in their cries that we see ourselves as part of nature, not separate from nature (emphasis in the original).[29]

For Senge, this shift represents a "structural" change, one that can make a difference in behavior.[30] For advocates of food system reform and environmental progress, it can serve as the ontological basis for sustainability.

Notes

Charles Brown made helpful suggestions on a previous version of this chapter.

1. For a useful summary of these issues I consulted Leo Horrigan, Robert S. Lawrence, and Polly Walker, "How Sustainable Agriculture Can Address the Environmental and Human Health Harms of Industrial Agriculture," *Environmental Health Perspectives* 110 (2002): 445–456.

2. See Wes Jackson and Jerry Glover, "The Need for a Taxonomy of Boundaries," in this volume.

3. The phrase belongs to Paul Ylvisaker and is cited in Virginia M. Esposito, ed., *Conscience and Community: The Legacy of Paul Ylvisaker*. New York: Peter Lang, 1999, 363.

4. This use of "amniotic" is borrowed from Edward Cohen, *The Peddler's Grandson: Growing Up Jewish in Mississippi*. New York: Dell, 2002.

5. See George Boody and Maria Krinke, *The Multiple Benefits of Agriculture: An Economic, Environmental and Social Analysis*. White Bear Lake, MN: Land Stewardship Project, 2002.

6. David Abram, *The Spell of the Sensuous*. New York: Vintage Books, 1997, 69.

7. Martin Heidegger, *Being and Time*, trans. J. Macquarrie and E. Robinson. New York: Harper and Row, 1962. In his later work, Heidegger analyzes the relationships among humans, technology, and nature in ways that appear sympathetic to various strains of environmentalism. But his full description of human–nonhuman boundaries, including those relevant to the natural world, is found in *Being and Time*. See *The Question Concerning Technology and Other Essays*, trans. W. Lovitt. New York: Harper and Row, 1977, 15; and *Discourse on Thinking*, trans. J. M. Anderson and E. H. Freund. New York: Harper and Row, 1966, 50. In the first reference, Heidegger notes that "Agriculture is now the mechanized food industry...driving on to the maximum yield at the minimum expense." In the second, he says that modern society has developed to the point where "Nature becomes a gigantic gasoline station, an energy source for modern technology and industry."

8. *Being and Time*, 65ff, 78ff, 107ff, 152ff.

9. Michael Zimmerman has done some of the best work on this issue. See "Rethinking the Heidegger-Deep ecology Environmentalism," *Environmental Ethics* 15 (1993): 195–224, and "Heidegger's Phenomenology and Contemporary Environmentalism," in *Eco-Phenomenology: Back to the Earth Itself*, Charles S. Brown and Ted Toadvine, eds. Albany: State University of New York Pres, 2003, 73–101. Also see Christian Diehm, "Natural Disasters," in *Eco-Phenomenology: Back to the Earth Itself*, 171–185.

10. *Being and Time*, 111.

11. Ibid., 465ff, 101.

12. Ibid., 440.

13. Ibid., 122–134, 413–414.

14. Ibid., 100.

15. Ibid.

16. Ibid. For an excellent description of the shift from the scientific to the aesthetic encounter with nature see Tom Stienstra, "Nature Can Show Up When You Least Expect It," *San Francisco Chronicle*, July 29, 2001, Sec. C, 18. The notion of something showing up captures a key Heideggerian feature of our engagement with the world. Stienstra's account sounds like the ontological basis for the contemporary "transformative" experience.

17. *Being and Time*, 165.

18. For example, see *Being and Time*, 98ff.

19. *The Economist*, October 30–November 5, 2004.

20. Ibid., 7–8.

21. Ibid., 8.

22. *New York Times*, November 22, 2004, E1.

23. Patrick Martins, "Introduction," in *Slow Food*, ed. Carlo Petrini. White River Junction, VT: Chelsea Green, 2001, xiv.

24. Hermann Scheer, "Region is Reason," in *Slow Food*.

25. "Radiation, Tobacco, and Illness in Point Hope, Alaska," in *The Environmental Justice Reader: Politics, Poetics and Pedagogy*, ed. Joni Adamson, Mei Mei Evans, and Rachel Stein. Tuscon: The University of Arizona Press, 105–124.

26. Ibid., 113.

27. Ibid., 118.

28. Ibid., 116.

29. *The Fifth Discipline*. London: Random House Business Books, 1999, 78.

30. Ibid., 53.

Chapter 13

Culture and Cultivation: Prolegomena to a Philosophy of Agriculture

Ted Toadvine

With few exceptions, academic philosophers have had little to say about agriculture, at least during the past 150 years or so. One such exception is a small book from the 1960s entitled *Roots in the Soil: An Introduction to Philosophy of Agriculture*, whose preface begins with these words: "One of the striking features of the history of philosophy...is the almost total absence of reflection on agriculture, agrarianism, and the significance of farm labor."[1] Paul Thompson, the author of another notable exception in the recent literature, puts the point more colorfully: "While sociology, economics, history, and literature tolerate rural studies, philosophy does not. Farming is like farting in most philosophical circles: one avoids mentioning it as assiduously as one avoids doing it."[2]

But, to be more precise, the recent avoidance of farming as a topic for philosophical inquiry is in particular an Anglo-American aversion. In the historical section of *The Second Sex*, French philosopher Simone de Beauvoir devotes an entire section to the "Early Tillers of the Soil," arguing that agriculture marks a turning point in masculine self-awareness and control over nature, leading to private property, cultural institutions, and a new temporal self-understanding.[3] Although she is rarely granted this honor, I believe these pages justify identifying Beauvoir as the first ecofeminist in our contemporary sense of the word.[4] According to Beauvoir, "all nature seemed to [the early agriculturalist] like a mother: the land is woman and in woman abide the same dark powers of the earth."[5] Although other thinkers have suggested that the original agricultural societies were matriarchal in character, Beauvoir denies any golden age of matriarchal power since, in her view, political power has always been masculine: "In spite of the fecund powers that pervade her, man remains woman's master as he is the master of the fertile earth; she is fated to be subjected, owned, exploited like the nature whose magical fertility she embodies."[6] Agriculture, then, is exploitation, perhaps in its most fundamental form; and this exploitation is symbolically and historically inseparable from the exploitation of women.[7] In Beauvoir's interpretation, agriculture marks the development of culture, while women find themselves on the side of nature, the Other.

We find the opposite interpretation of agriculture just a few years later in Heidegger's "The Question Concerning Technology":

> The field that the peasant formerly cultivated and set in order appears differently than it did when to set in order still meant to take care of and maintain. The work of the peasant does not challenge the soil of the field. In sowing grain it places seed in the keeping of the forces of growth and watches over its increase. But meanwhile even the cultivation of the field has come under the grip of another kind of setting-in-order, which *sets upon* nature. It sets upon it in the sense of challenging it. Agriculture is now the mechanized food industry.[8]

For Heidegger, the peasant farmer is on the side of nature; he does not *challenge* nature but works together with it. The break from nature is a function of the setting-in-order, the enframing of nature apparent in contemporary agribusiness. We see here the agrarian strain of thought in Heidegger, along with its politically reactionary overtones.[9]

Beauvoir and Heidegger locate agriculture at different points on the nature–culture spectrum. Or, rather, agriculture lies at the juncture of nature and culture, but, for each, it faces the opposite direction. This ambivalence is apparent in the shared etymological origin of our terms "culture" and "cultivation." Both have their root in the Latin *cultus*, "tilled or cultivated," past participle of *colō/colēre*, to cultivate or till the soil. Although the primary sense of the verb was agricultural, its derivative senses are also relevant: to *inhabit* [a place], to *practice or cultivate* [one's pursuits], or to *worship* [the gods]. These different senses are reflected in the derived noun, *cultus*, referring not only to cultivation in the agricultural sense, but also to training, style of dress, refinement, sophistication, and civilization. Except for the religious sense, culture as worship, which became obsolete in English after the fifteenth-century (but remains in our word "cult"), our current terms "culture" and "cultivation" retain these many figurative echoes. They also clearly retain the interesting double sense or ambiguity of the original: On the one hand, cultivation is the dirty work of tilling the soil. This is what those sweaty farmers do out in the fields, outside of town, far from the opera house and the fine dining establishments. On the other hand, "cultivation" is sophistication, refinement, gentility, keeping your hands clean, pursuing the life of the mind, acting civilized. The terms "culture" and "cultivation" harbor the seed of an entire series of oppositions like mind versus body, city versus country, theory versus practice, culture versus nature. *Cultus*, our Latin starting-point, is related to *culter*, blade, as in the blade of the plow. This is clearly a blade that cuts two ways.

Agriculture as Nature-Culture Boundary

This strange ambivalence of "culture/cultivation" results from its situation as the fundamental boundary, the mediating point or site of passage, between nature and culture. In justifying their call for a philosophy of agriculture, Hill and Stuermann make this point:

> the complex superstructure of a sophisticated, technological civilization rests upon that group of workers who handle the soil and deal with nature's resources and who are, therefore, an indispensable link between the level of industrialized, urban civilization and the riches and resources of nature from which humane culture springs.[10]

This position is stated more plainly by classical scholar Stephanie Nelson in her examination of the metaphysics of farming in Hesiod and Vergil: "To discuss our relation to 'nature' is to discuss the interconnection of society, wild nature, and domestic nature. It is, in other words, to discuss farming, since farming is where nature and culture meet."[11]

Seen from the side of nature, cultivation is the first and essential step toward civilization, the fundamental human manipulation of nature that makes all later technological and social development possible. But seen from the side of "culture," the farmer is on the outside, out in the natural world. Leaving the city for the farm is a "return to nature," to a "natural" way of life. Agriculture is at the edge—the margin, the barbarian frontier—of culture. This Janus-headed quality is a consequence of farming's situation as a fundamental boundary or threshold. In a sense, agriculture is the excluded center of culture, the supplement that founds the system, the outside that makes the constitution of the inside possible.

While, from the perspective of culture, agriculture is the Other, the excluded, it is simultaneously the condition for the possibility of civilization as we know it. This is perhaps a point that seems too obvious to warrant mention. "First, there must be food," Aristotle tells us when enumerating the functions of a state, and thus there must be farmers to procure it.[12] Rousseau writes that, "for the philosopher, it is iron and wheat that have civilized men and ruined the human race."[13] Agriculture, for Rousseau, makes possible the first large-scale cooperative efforts of human technology and the first division of labor. This moment marks the introduction of inequality into nature, since it leads with inevitability toward the division and accumulation of property. While Rousseau's speculations often fall short of anthropological accuracy, he is not alone in identifying agriculture as the crucial step that marks the historical and technological dawn of our culture, nor alone in condemning this beginning. For Daniel Quinn, whose recent string of ecologically minded "novels" has attracted a cult following, the neolithic agricultural revolution represents the parting of ways in human societies: while the "leavers," the world's vanishing hunter-gatherer societies, continue the traditions of human life handed down through millions of years of cultural acquisition, the "taker" society, having crossed the spiritual and mental threshold of the agricultural revolution, practice a massive cultural amnesia designed to obliterate any recollection of pre-agricultural human life. The transition to "taker" life represents, for Quinn, the crossing of a fundamental spiritual and mental "border."[14]

It is worth emphasizing that the question of agriculture is far from academic. Even if the majority of academic philosophers have avoided the question of agriculture altogether, other ecologically minded thinkers have equated the "agricultural revolution" with our civilization's fall from grace. As we cast our glance back at the dawn of our culture, the Czech phenomenologist Erazim Kohák notes, "the transition to the pastoral/agricultural mode of sustenance might well seem like the original sin, a step from the Garden of Eden directly to Broadway and 42nd Street at 1:00 a.m. on a hot Friday night."[15] Daniel Quinn, in fact, takes the association of the agricultural revolution with the story of the Fall quite literally, reading the Cain and Abel story as a piece of ancient war propaganda. On his reading, the murder of Abel symbolizes the slaughter of Semitic herders by Caucasian agriculturalists expanding their territory.[16] To non-agriculturalists, Quinn suggests, the agricultural way of life must have seemed like a curse, like a punishment for the original sin. This original sin is our culture's claim to have the knowledge of the gods, to be in a position to decide who should eat and who should starve, who should live and who should die. The curse of Cain is the curse of "taker" culture, the curse of agriculture, our curse. Whereas the rest of the natural community accepts, by default, a law of limited competition, the underlying premise of our agricultural practices is to wage total war on any creature that resists our aspiration to reproduce without limit. We carry out this war of "totalitarian agriculture" by annihilating our competitors, destroying their food, and denying them access to food, with the end goal of converting all resources available on the earth to the production of food for humans.[17] The same desire for absolute control over our own destinies that drove the neolithic revolution should be recognizable as the underlying motivation for our entire technological civilization.

The philosophical problem of agriculture is not a technical problem; it is distinct from such questions as whether we can feed all of the starving people in the world, or whether we can make our agricultural practices sustainable while continuing to maximize production. The question concerns instead whether such technological progress would count, in a deeper sense, as real success—whether, as Paul Shepard contends, the agricultural way of life would remain only the "next-to-best of all possible worlds."[18] The philosophical problem of agriculture cannot be solved by developing new technological fixes, since it concerns the meaning of the agricultural way of life, the relationship between agriculture and human self-understanding, and the relationship between nature and culture.

Risk and Faith in the Seed

The first piece in the puzzle of the meaning of agriculture is the seed. Let us return to Heidegger's description of the peasant farmer: "In sowing grain it [i.e., the work of the peasant] places seed in the keeping of the forces of growth and watches over its increase."[19] The characteristic activity of the farmer is, for Heidegger, the sowing of seed that places it in the earth's keeping. Heidegger emphasizes the element of faith and trust between peasant and earth, to be contrasted in his account with the control exercised by contemporary mechanized agriculture.

The other side of the peasant's trust of the seed to the earth, as Rousseau notes, is risk:

> to devote oneself to [large-scale cultivation] and to sow the lands, one must be resolved to lose something at first in order to gain a great deal later: a precaution quite removed from the mind of the savage man, who, as I have said, finds it quite difficult to give thought in the morning to what he will need at night.[20]

The sowing of grain is a gamble with precious food, the sacrifice of current sustenance for the prospect of long-term return. If, as Freud has suggested, the development of culture is marked by the renunciation of an instinct or initial impulse, we can see the planting of grain as a significant moment of such renunciation.[21] The saving of grain through a hungry winter would involve an appreciable degree of restraint and short-term sacrifice, and the repetition of the cycle depends utterly on a sufficient annual harvest.

Many authors have also shared Rousseau's conviction that the planting of seed marks a shift in human temporal orientation, opening as it does a future-oriented world in which investment and trust take on new meaning.[22] Planting entails permanence of location (or at least a determined route of migration), and this permanence subtends the unity of the clan with a particular land. As Beauvoir writes, "In place of the outlook of the nomadic tribes, living only for the moment, the agricultural community substituted the concept of a life rooted in the past and connected with the future."[23] The permanence of the land provides the temporal continuity of past and present; the seed expresses our relation to the future.[24]

The sowing of seed, then, involves sacrifice and renunciation, as well as a gamble on the future.[25] The correlate of this risk is faith in the earth as a partner in production. Kohák speaks of the herder/farmer's relationship with the earth as a relationship of "community," according to which the earth as partner is treated as our "intimately known kin."[26] Although early agriculturalists did not see the earth as divine, as had their hunter/gatherer predecessors, they did interact with it as an animate, living organism toward which proper care and stewardship should be shown.

Presented in this fashion, the seed represents a relation of faith and trust in the earth. But perhaps we have this completely inverted; perhaps we should interpret the sowing of seed as *precisely the opposite* of faith and trust. What is the storing and sowing of seed if not a profound *distrust* in the provision of the earth for our well-being? Is not the planting of seed precisely our attempt to wrest control of our own destinies from "the hands of the gods," as Quinn has described it? For Paul Shepard, the shift to agrarian thought from hunting and gathering marks the transition "from a core process of chance to one of manipulation, from reading one's state of grace in terms of the success of the hunt to bartering for it, from finding to making, from sacrament received to negotiations with humanlike deities."[27] The planting of the seed, then, would be less a sacrificial gift to the earth than an investment to hedge one's bets, an exchange in an economy of barter with the gods and the earth. If, as Heidegger teaches us, the opening of the future is a relation to one's own death, the storing up of grain is a technique to fend off finitude. The seed represents not only the future, but also sacrifice: it is a small death to stave off the larger death that comes toward us from the future. The meaning of the seed, and by extension of seed-based agriculture, is tied up with our relationship to death.

According to Hans Jonas, death is the first primordial mystery to confront our animistic human ancestors. If the world in toto is characterized by life, the one riddle that resists solution is death: it is, Jonas writes, "the contradiction to the one intelligible, self-explaining, 'natural' condition which is the general life."[28] The first solution to this contradiction, the animist solution, is a denial of death by transmuting it into life itself; death is only life postponed or transfigured. The ambiguity of this "solution" is apparent in the phenomenon of early tombs, which both mark the recognition of death and imply a life beyond it. Is the "planting" of the deceased body, like that of the seed, perhaps connected with the expectation of its resurrection?

According to Joseph Campbell, seed-based agriculture offered our ancestors new structures for mediating their relationship with death. By borrowing structures from the life of plants—going to seed, sowing, rebirth—our mythology develops a structure for reconciling the finite and the infinite, namely, the structure of sacrifice. Life encompasses death and requires it as an essential moment. The mythology of early agriculturalists is rife with deities whose self-sacrifice bring life, often in the form of the first domestic plants. Human sacrifices, real and symbolic, function as repetitions of this cosmic event that maintain the earth's fertility. Campbell cites the work of Adolf Jensen, according to whom human sacrifices in fertility rituals are "but the renditions in act of a mythology inspired by the model of death and life in the plant world":[29]

> That killing should have assumed such a prominent position in the total view of the world in this culture sphere, I should like to refer quite specifically to the occupation of these people with the world of plants. There was here revealed to mankind, in some measure, a new field of illumination. For the plants were continually being killed through the gathering of their fruits, yet the death was extraordinarily quickly overcome by their new life. Thus there was made available to man a synthesizing insight, relating his own destiny to that of the animals, the plants, and the moon.[30]

The cycle of planting and rebirth gives rise to a mythology of planting and rebirth among early agricultural societies. Religions of sacrifice follow the model of the seed and appeal to the plant aspect of our own inner natures. The sacrifice, as Deleuze and Guattari say of drunkenness, is "the triumphant irruption of the plant within us."[31] But it is not the sacrifice of the individual to the whole that makes us drunk, as Campbell notes, but rather the destruction inherent in life's own creativity:

> where the silent cosmic feeding of the plant kingdom on the substances of its mother Earth is the one, ever-present, inescapable, immediate experience of the meaning of life in a living environment, there is always a moment of ultimate frightfulness, executed either in some hidden place, or in full view, in the pivotal climax of an enveloping, dreamlike revel of feasting and hilarity, masks, music and dance, in celebration of the sublime frenzy of this life which is rooted (if one is to see and speak truth) in a cannibal nightmare.[32]

So far, we have traced the meaning of agriculture to the seed, and our encounter with the seed is a means of working out our relationship to the death that comes to us from the future. The working-out of death involves a certain "becoming plant" of the early agriculturalists, and we have so far characterized this becoming plant in two seemingly divergent ways: as faith and trust, on the one hand, and as sacrifice and frenzy, on the other. What is the relationship between these two ways of becoming plant, and what relationship with death do they express?

Becoming Plant

For a first pass at this problem, let us consider a contrast between the two "orders of life," plant and animal, that Campbell borrows from Oswald Spengler. The way of the plant is characterized by rootedness in the earth, bondage to the whole. Lacking will, choice, or individuality, the plant exists only as part of the larger cycle of life that moves through it, within and without. The animal, by contrast, is free, individuated, independent. While the life of the plant is cosmic, that of the animal is microcosmic, a little world within a larger world, an organism relatively distinct from its *Umwelt*: "Only insofar as a living unit can distinguish itself from the All and determine its position with respect to it, can it be said to be a microcosm."[33] While the plant is characterized by rhythmic periodicity, Spengler writes, the animal exhibits the tension of polarity:

> All that is cosmic is characterized by periodicity, it has a pulse, a rhythm. All that is microcosmic is distinguished by polarity: the word "against" epitomizes its whole character. It possesses tension. We speak of tense alertness, intense thought. But *all* awake relationships are by nature tensions: sense organs and perceived object, I and Thou, cause and effect, subject and predicate. Each of these implies a distinction in space and thereby a "tension" [*Spannung*], and where the state that is significantly called *détente* [*Abspannung*, "relaxation"] appears, fatigue immediately sets in and finally, in the microcosmic system, sleep. A human being asleep, relieved of every tension, is in a plantlike state.[34]

In Campbell's interpretation, the animal and the plant orders operate as conflicting urges within human consciousness, and different human cultures, as demonstrated in their myths and rites, resolve this conflict in favor of one or the other mode of existence. Communities centered on the hunt take the carnivore as their model, resolving the relation with death through rituals of slaughter and reconciliation that build tension rather than relaxing it. The will in nature, for a *Spannung* society, is represented by the powers of the animal. The agricultural life of *Abspannung*, on the other hand, models itself on the way of the plant: rooted, rhythmic, communal, torporous. Becoming plant is becoming one, losing individuality, being absorbed into cosmic life, and this historically unfolds in one of two directions: either through unity with a transcendent reality (e.g., Christianity) or by realizing one's own metaphysical identity as a microcosm within a macrocosm (e.g., Buddhism).[35] Although these two directions are distinct and one will always be ascendent, we should note that human societies will inevitably include both elements: the animal, both human and nonhuman, is polar, since it includes the plant way within itself. Therefore, animal life, including that of the human animal, introduces a primordial division or fold in being. While, at the plant level, it strives for return to the security of fixed being, it is at the same time, as animal, inescapably free, thrown into the world of choice.

As a modification to this model, let us suggest first of all that the identification of animals with free individuality and plants with bondage to the earth is simplistic, since freedom does not first arise with the animal, but rather, as Hans Jonas has suggested, marks the boundary between inorganic and organic. Life is freedom, since the first form of freedom is metabolism itself.[36] Life segregates itself from physical matter by introducing into being a fundamental virtuality or possibility, a "to be or not to be," the possibility of death. Threatened by nonexistence, life responds by affirming existence, and its own being is therefore at issue for it from the very beginning. In life as such, being implies the possibility of nonbeing, and therefore life essentially implies death and constant crisis.[37] The irruption of the organic introduces into reality the most fundamental form of freedom. Plant, animal, and human are, for Jonas, relative forms the development of this freedom may take.

Nor can we, strictly speaking, identify the animal with individuality and the plant with collectivity. First, there are animals that follow the communal and frenzied model, for example, pack animals: wolves, rats, piranha. The behavior of the pack is something other than the behavior of freely choosing and independent individuals, while it is also, at the same time, something other than a "common sense" that would unite the organs as simple parts of one whole. On the other hand, the "unity" or holism of plant life is more complex than has been represented. There are, in the first place, relatively individuated plants: seed plants and trees, for instance, that function relatively as individual units even while bound to a single location. But there are also grasses and weeds that defy the very dichotomy of individual verses group. When a grass plant sends out multiple rhizomes that develop into semiautonomous offshoots, capable of surviving independently but having the same genetic makeup, are we talking about one plant or many?[38] Deleuze and Guattari give the name *rhizomes* to such multiplicities—neither single nor multiple—whether animal or plant: "The rhizome itself assumes very diverse forms, from ramified surface extension in all directions to concretion into bulbs and tubers. When rats swarm over each other. The rhizome includes the best and the worst: potato and couchgrass, or the weed."[39]

Freedom, as Jonas intends it in distinguishing organic from inorganic, is one form of what Deleuze and Guattari call "deterritorialization," the opening up for a rhizomatic system of a line of flight, a new trajectory, an escape route. Rather than associating plants with bondage and animals with freedom, we must instead admit the becoming-animal of the plant and the becoming-plant of the animal:

> The orchid deterritorializes by forming an image, a tracing of a wasp; but the wasp reterritorializes on that image. The wasp is nevertheless deterritorialized, becoming a piece in the orchid's reproductive apparatus. But it reterritorializes the orchid by transporting its pollen. Wasp and orchid, as heterogenous elements, form a rhizome...a veritable becoming, a becoming-wasp of the orchid and a becoming-orchid of the wasp.[40]

Alongside the way of the plant and the way of the animal, we must recognize the way of the rhizome, which is the way of multiplicity, a singularity that is neither one nor many.

To understand the ramifications of this alternative, let us sort out the threads that have brought us to this point. We began with the seed as our guiding clue to the meaning of the agricultural life, and we saw that the meaning of the seed is bound up with a resolution of the relation to death. The way of the seed is that of sacrifice and resurrection: the material body is "planted" in order to make possible the resurrection of the soul. In other words, the sprouting seed becomes the sign of the cross.[41] The seed, for instance a mustard seed, becomes the symbol for faith in the afterlife, in the definitive transcendence of the cycle of birth and death. Once the seed becomes the soul, and the body becomes the tomb, we have achieved a reversal of the original panvitalism for which death served as ultimate contradiction. Now death is ubiquitous, and it is life that requires super-natural explanation.[42] In the terminology of Deleuze and Guattari, we could say that, within the "assemblage" of the early agricultural world, the seed "expresses" transcendence, the passage beyond the limit.[43] It is this basic passage beyond the limit that makes the transition to a dualistic view of the natural world possible.

All of Western thought—from taxonomy to linguistics, politics to theology, ontology to anatomy—has been dominated, Deleuze and Guattari have claimed, by the structure of the tree: hierarchical, linear, foundational, symmetrical, teleological, well grounded. Even our seed-based agriculture follows this pattern, in contrast with the rhizomatic orientation of the East.[44] The faith of the rhizome is not the faith in transcendence, not the gamble of everything on a single toss, "to be or not to be." It is, rather, faith in the multiple, in the duplication of possibilities, which situates the rhizome always at the boundary, at the "in between."[45] The contrast with the order of the seed, then, is the creative flight of the rhizome, of a nonhierarchical, nonteleological multiplicity, or, in other words, the order of the weed. Western thought and culture have been haunted by the weed in many forms: rats, criminals, subversives, terrorists, cockroaches, crabgrass.[46] The exclusion of the weed within our agricultural practices and thinking is perhaps the obverse of our cultural orientation toward transcendence and purification from nature. If agriculture is the pivot point or hinge connecting nature and culture, then reworking the relation between humans and nature requires an agriculture oriented around the weed rather than the seed, such as we find it, for instance, among the Tepehuan people of Nabogame.[47] The Tepehuan encourage the introgression of weedy, ancient teosinte in their fields of domesticated maize, leading to gene "leakage" back and forth between the varieties that alters both without integrating them into a single strain. Indigenous cultural practices, domesticated plants, and native "weeds" enter here into a single block of mutually transformative coding and recoding, a veritable becoming-other of each of the terms involved. Such weedy agriculture provides a figure for rethinking human-nature relations as a system of mutual exchange and becoming; it is just as much the plant, wild and domesticated, that farms itself through us as we who farm it.

Seen in this light, the problem of agriculture rejoins the larger debate among environmental theorists concerning the "place" of humans within nature. In the rhetoric of Rousseau, Shepard, and Quinn, agriculture marks a "fall" from "nature" into "culture." But what do these authors understand by "nature" and "culture," and what ideal relations do they envision between these terms? William Cronon finds in such condemnations of agriculture the traces of what he has called the "wilderness myth": "nature, to be natural, must also be pristine—remote from humanity and untouched by our common past."[48] As illustration of the wilderness myth at work, Cronon cites remarks about agriculture by Earth First! Founder Dave Foreman:

> Before agriculture was midwifed in the Middle East, humans were in the wilderness. We had no concept of "wilderness" because everything was wilderness and *we were a part of it*. But with irrigation ditches, crop surpluses, and permanent villages, we became *apart from* the natural world.... Between the wilderness that created us and the civilization created by us grew an ever-widening rift.[49]

The divorce of humans from nature that Foreman attributes to agriculture leaves no possibilities for reunion—not unless we wish, as Cronon puts it, to "follow the hunter-gatherers back into a wilderness Eden and abandon virtually everything that civilization has given us."[50] From a practical perspective, such a position offers no more than a "self-defeating counsel of despair," according to Cronon: "if nature dies because we enter it, then the only way to save nature is to kill ourselves."[51] The defeatism of this position stems directly from the exclusive disjunction it establishes between nature and culture. "Real" nature excludes human culture, and consequently agriculture, by definition; on the other hand, "civilized" human beings and their activities are no longer a genuine part of the natural world. Black-and-white metaphysical categories here come to the service of a political agenda, but what is the rationale for cutting reality at these joints?

In taking the ideal of "pure" nature as our standard, Cronon suggests, we turn a blind eye to the environmentally harmful practices of our everyday lives, the "middle ground" where we actually live. We also denigrate the lives and work of those people who actually make a living from the land, whether indigenous inhabitants or "country people" who "generally know far too much about working the land to regard *unworked* land as their ideal."[52] Antiagriculturalists, by ignoring the actual experience of farmers, have missed those aspects of nature that may be disclosed only through such experience, such as the love for and knowledge of a particular place that Wendell Berry finds at the heart of stewardship for the land, or the sacramental experience of indigenous foods and the mutualistic role of humans as pollinators and seed dispersers described by Gary Paul Nabhan. What we make of such examples depends in large part on whether we leave open the possibility of a fruitful interaction between culture and nature, humans and the land. Even if there are genuine differences between the processes of spontaneous nature and those that are guided by human interests and interactions, agriculture marks a threshold where a new hybrid may emerge, where we may perhaps be able to speak of a rhizomatic co-evolution that is no more ours than nature's own. It is toward an investigation of such mutual exchange that a philosophy of agriculture must turn, which gives it a privileged position in articulating a new language for that worked ground where nature and culture meet.

Notes

1. Johnson D. Hill and Walter E. Stuermann, *Roots in the Soil: An Introduction to Philosophy of Agriculture*. New York: Philosophical Library, 1964, ix.

2. Paul B. Thompson, "Agrarianism as Philosophy," in *The Agrarian Roots of Pragmatism*, ed. Paul B. Thompson and Thomas C. Hilde. Nashville: Vanderbilt University Press, 2000, 49.

3. Simone de Beauvoir, *The Second Sex*, trans. H. M. Parshley. New York: Vintage Books, 1989, 66–81.

4. See, for instance, Karen Warren's use of this concept in "The Power and the Promise of Ecological Feminism," *Ecological Feminist Philosophies*, ed. Karen Warren. Bloomington: Indiana University Press, 1996, 19 and passim.

5. Beauvoir, *The Second Sex*, 68.

6. Ibid., 73.

7. Echoes of Beauvoir's position may be found in Paul Shepard's *Coming Home to the Pleistocene*. Washington, DC: Island Press, 1998, e.g., 71–75, 95–97.

8. Heidegger, "The Question Concerning Technology," in *Basic Writings*, revised edition, ed. David Farrell Krell. San Francisco: Harper, 1993, 320.

9. Thompson suggests that the historical tie of agrarian thought with blood-and-soil nationalism, such as we find it explicitly expressed in German National Socialism, takes us part of the way toward understanding the absence of any serious interest in agrarian thought among contemporary theoreticians: "Saying anything that does not connote immediate disapproval of agrarian themes is quickly taken to constitute evidence of racism and insensitivity on the part of the speaker" (Thompson, "Agrarianism as Philosophy," 49).

10. *Roots in the Soil*, xi.

11. Stephanie Nelson, *God and the Land: The Metaphysics of Farming in Hesiod and Vergil*. New York: Oxford University Press, 1998, 152.

12. *Politics*, Book VII, Ch. 9, 1328b6.

13. *Discourse on the Origin of Inequality*, trans. Donald Cress. Indianapolis: Hackett, 1992, 51–52.

14. See Daniel Quinn, *Ishmael*. New York: Bantam Books, 1992, especially chapters 8 and 9; Quinn, *The Story of B*. New York: Bantam Books, 1996. On the notion of the "border" or "boundary," see *The Story of B*, 128. Although Quinn sees the agricultural revolution as the dawn of our problematic worldview, he does not rule out the possibility of an ecologically friendly agriculture. On this point, his work stands opposite that of Paul Shepard, for whom an agriculture-based way of life remains inherently second-best.

15. Kohák, "Varieties of Ecological Experience," *Environmental Ethics* (Summer 1997): 158.

16. Shepard suggests that war as such is a consequence of the agricultural mind set: "War and warriorhood probably grew out of the territorialism inherent in agriculture and its exclusionary attitudes and the necessity for expansion because of the decline of field fertility and the frictions and competitions of increased human density" (*Coming Home to the Pleistocene*, 86).

17. In *The Story of B*, Quinn is very clear that his criticisms are aimed only at what he terms "totalitarian agriculture," not agricultural practices that avoid this war-waging agenda. See especially 151–155, 247–260.

18. This is Paul Shepard's characterization of the agrarian solutions proposed by Gary Snyder, Wes Jackson, and Wendell Berry. See *Coming Home to the Pleistocene*, 107n50.

19. Heidegger, "The Question Concerning Technology," 320.

20. *Discourse on the Origin of Inequality*, 52.

21. Freud, *Civilization and its Discontents*, trans. James Strachey. New York: Norton, 1961, 43n. Freud's own discussion concerns the renunciation of the urge to urinate on fire, which would mark an earlier moment in the attainment of civilization, but the parallel is easily drawn with the renunciation of the desire to eat gathered food.

22. Note also Stephanie Nelson's discussion of the "essential ambiguity" of the moment of sowing in Hesiod's *Works and Days*, which is the paradigmatic moment for conveying Hesiod's lesson of "due season." See *God and the Land*, chapter 6, especially 168–169.

23. *The Second Sex*, 67.

24. Nevertheless, it would be a mistake to conclude that the human relation with the future opens for the first time with the planting of seed. Philosophers from Plato to Levinas have emphasized the relation to the future that holds through one's progeny, and, for women at least, this relation to the future would necessarily have preceded agriculture. Note, in this regard, Engels's controversial claim that monogamy develops in parallel with private property, both on the backs of agriculture, as a means for male property owners to guarantee the patrimony of their successors. See Frederick Engels, *The Origin of the Family, Private Property, and the State*. New York: International Publishers, 1972, 128–129.

The narrative capacity at work in the cooperative hunt also apparently implies a temporal sense beyond that achievable by nonhuman animals. Quinn provides a provocative argument for this point in *The Story of B*, 164–176.

25. I have developed this interpretation of the planting of seed in more detail in my essay, "Ecophenomenology in the New Millennium," in *The Reach of Reflection: Issues in Phenomenology's Second Century*, ed. Steven Crowell, Lester Embree, and Samuel Julian. Boca Raton: Center for Advanced Research in Phenomenology, 2001.

26. "Varieties of Ecological Experience," 157–158.

27. *Coming Home to the Pleistocene*, 81.

28. Jonas, *The Phenomenon of Life: Toward a Philosophical Biology*. Evanston: Northwestern University Press, 1966, 8.

29. Campbell, *The Masks of God: Primitive Mythology*. New York: Viking Press, 1959, 171.

30. Adolf Jensen, *Das religiöse Weltbild einer frühen Kultur*. Stuttgart: August Schröder Verlag, 1949, 168–170. Cited (in translation) by Campbell, 178.

31. Deleuze and Guattari, *A Thousand Plateaus*. Minneapolis: University of Minnesota Press, 1987, 11.

32. Campbell, *Historical Atlas of World Mythology*, Vol. II: The Way of the Seeded Earth, Part 1: The Sacrifice. New York: Harper & Row, 1988, 44.

33. Oswald Spengler, *Der Untergang des Abendlandes*. Munich: C.H. Beck'sche Verlagsbuchlandlung, 1930, vol. 2, 4–5. Cited in Campbell, *Historical Atlas of World Mythology*, Vol. II, Part 1, 32. See also pp. 8–9.

34. Spengler, 4–5; cited in Campbell, *Historical Atlas of World Mythology*, Vol. II, Part 1, 32–33.

35. See Campbell, *Historical Atlas of World Mythology*, Vol. II, Part 1, 8–11, 32–33, 44, and passim.

36. Jonas, *The Phenomenon of Life*, 3.

37. This constant crisis of life, for Jonas, also implies faith: "Committed to itself, put at the mercy of its own performance, life must depend for it on conditions over which it has no control and which may deny themselves at any time. Thus dependent on propitiousness or unpropituousness of outer reality, it is exposed to the world from which it has seceded, and by which it must yet maintain itself" (Jonas, 5).

38. The rhizomatic plant fulfills the human desire for repetition through posterity in the child, since the offspring *is* the parent.

39. Deleuze and Guattari, *A Thousand Plateaus*, 7.

40. Ibid., 10.

41. Given what I will say later concerning the chiasmatic character of the rhizome, I would venture this claim: the cross is a reterritorialized chiasm, just as faith in a transcendent reality is a reterritorialized perceptual faith.

42. I have in mind the reversal described by Jonas in "Life, Death, and the Body in the Theory of Being," *The Theory of Life*, 7–37.

43. I am thinking especially of the account of these terms offered in the fourth plateau, "November 20, 1923—Postulates of Linguistics," of *A Thousand Plateaus*, 75–110.

44. On this point, see their references to the work of André Haudricourt (*A Thousand Plateaus*, 18).

45. "The tree imposes the verb 'to be,' but the fabric of the rhizome is the conjunction, 'and...and...and...'. This conjunction carries enough force to shake and uproot the verb 'to be'" (Deleuze and Guattari, *A Thousand Plateaus*, 25. See also 98f.); and "AND is neither one thing nor the other, it's always in between, between two things; it's the borderline, there's always a border, a line of flight or flow, only we don't see it, because it's the least perceptible of things. And yet it's along this line of flight that things come to pass, becomings evolve, revolutions take shape" (Deleuze, *Negotiations*, trans. Martin Joughin. New York: Columbia University Press, 1990, 45).

46. As Henry Miller writes, "The weed is the Nemesis of human behavior.... Of all the imaginary existences we attribute to plant, beast and star the weed leads the most satisfactory life of all.... Eventually the weed gets the upper hand.... The weed exists only to fill the waste spaces left by cultivated areas. *It grows between*, among

other things. The lily is beautiful, the cabbage is provender, the poppy is maddening—but the weed is rank growth...it points a moral" (Henry Miller, in Henry Miller and Michael Fraenkel, *Hamlet*. New York: Carrefour, 1939, 105–106. Cited in Deleuze and Guattari, *A Thousand Plateaus*, 18–19).

47. See Gary Paul Nabhan's discussion of Tepehuan practices and other "weedy" forms of agriculture in "Fields infused with Wildness," chapter three of *Enduring Seeds: Native American Agriculture and Wild Plant Conservation*. San Francisco: North Point Press, 1989, 31–45. My thanks to Janet Fiskio for bringing this reference to my attention.

48. William Cronon, "The Trouble with Wilderness; or, "Getting Back to the Wrong Nature," in *Uncommon Ground: Rethinking the Human Place in Nature*, ed. By William Cronon. New York: W. W. Norton, 1996, 83.

49. Dave Foreman, *Confessions of an Eco-Warrior*. New York: Harmony Books, 1991, 69; cited in Cronon, "The Trouble with Wilderness," 83.

50. Cronon, 83–84.

51. Ibid., 83.

52. Ibid., 79.

Contributors

T. Clay Arnold is Professor of Political Science at the University of Central Arkansas. His scholarship features contemporary social and political theory and the water politics of the arid American West. He is the author of "Rethinking Moral Economy" (*American Political Science Review*, March 2001), *Thought and Deeds: Language and the Practice of Political Theory* and *An Introduction to Politics and Political Inquiry*.

Charles S. Brown, Professor of Philosophy at Emporia State University in Emporia, Kansas, is co-editor of *Eco-Phenomenology: Back to the Earth Itself*, and author of several essays and book chapters on the convergence of phenomenology and environmental thought.

J. Baird Callicott, Professor of Philosophy at University of North Texas, is author and editor of many books, including *Beyond the Land Ethic: More Essays in Environmental Philosophy*, *Earth's Insights: A Survey of Ecological Ethics from the Mediterranean Basin to the Australian Outback*, and *In Defense of the Land Ethic: Essays in Environmental Philosophy*.

Beth Dempster is an associate of the Civics Research Co-operative, a not-for-profit co-op focused on equity, sustainability, and citizen involvement. As a Ph.D. candidate in Planning at the University of Waterloo, she is writing a hypertext dissertation to integrate sustainability, complex systems, and praxis: www.sympoiesis.net. She has published papers in *Environments* and has presented at conferences of the International Society for Systems Science, the Environmental Studies Association of Canada, and the International Society for Ecological Economics, among others.

Strachan Donnelley founded the Center for Humans and Nature in 2003 and serves as its president. Among numerous published articles in philosophy and applied ethics, Donnelley has co-edited and written for three Special Supplements to the Hastings Center Report: "Animals, Science, and Ethics," "The Brave New World of Animal Biotechnology," and "Nature, Polis,

Ethics: Chicago Regional Planning." He also edited a special edition on the philosopher and ethicist Hans Jonas, also in the Hastings Center Report. Recently, he has written several articles on philosophy, evolutionary biology, and ethical responsibility, including work on Whitehead, Jonas, Ernst Mayr, and Aldo Leopold.

Jerry Glover, Research Scientist in Natural Systems Agriculture at the Land Institute, received his Ph.D. in Soil Science from Washington State University in 2001. His graduate work involved assessing the impacts of apple production systems on soil, crop, and environmental quality, disease and pest management, and financial performance.

Bruce Hirsch is executive director of the Clarence E. Heller Charitable Foundation in San Francisco. He is responsible for the foundation's grant-making programs in environment and human health and the sustainable management of resources.

Wes Jackson, President of the Land Institute in Salina, Kansas, holds a Ph.D. in genetics from North Carolina State University. He established and served as chair of one of the country's first environmental studies programs at California State University–Sacramento and then returned to his native Kansas to found the Land Institute in 1976. He is the author of several books including *New Roots for Agriculture* and *Becoming Native to This Place* and is widely recognized as a leader in the international movement for a more sustainable agriculture. He was a 1990 Pew Conservation Scholar, a 1992 MacArthur Fellow, and received the Right Livelihood Award (the "alternative Nobel prize") in 2000.

Jon Jensen directs the Environmental Studies program and teaches philosophy at Luther College in Decorah, Iowa. Jensen received his Ph.D. in philosophy from the University of Colorado and taught in Vermont before returning to the Midwest where he has deep roots. He is the co-author of *Questions That Matter: An Invitation to Philosophy* as well as various essays. His current research focuses on connections between sustainable agriculture, ecological restoration, and local food systems.

Irene J. Klaver, Director of the Philosophy of Water Project and Assistant Professor of Philosophy at the University of North Texas, has authored numerous book chapters as well as articles in *Environmental Ethics*, *Research in Phenomenology*, *KRISIS: Tijdschrift voor Empirische Filosofie*. She is co-editor of the widely used textbook *Environmental Philosophy: From Animal Rights to*

Social Ecology. She has a book forthcoming on boundaries, called *Of(f) Limits—Environmental Philosophy in a Continental Mode*.

Max Oelschlaeger, Frances B. McAllister Chair of Community, Culture, and Environment at Northern Arizona University, is author and editor of many books, including *Caring for Creation: An Ecumenical Approach to the Environmental Crisis*, *The Idea of Wilderness: From Prehistory to the Age of Ecology*, and *The Environmental Imperative: A Socio-Economic Perspective*.

Firooza Pavri is Assistant Professor of Geography at the University of Southern Maine's Geography-Anthropology Department. Her research interests include development studies, human–environment interaction, and natural resource management. She has published in journals such as *Physical Geography*, *Geoforum*, and the *International Journal of Population Geography*. She has also been the recipient of several grants and most recently was funded through NASA for research on monitoring rural resource trends using satellite imagery.

Anna L. Peterson is Professor in the Department of Religion and an affiliate of the Center for Latin American Studies and the School of Natural Resources and the Environment at the University of Florida. She is author of *Being Human: Ethics, Environment, and Our Place in the World* and *Martyrdom and the Politics of Religion*, and co-editor of *Christianity, Globalization, and Social Change in the Americas*. Her most recent book is *Seeds of the Kingdom: Utopian Communities in the Americas*.

Ted Toadvine is Assistant Professor of Philosophy and Environmental Studies at the University of Oregon. His interests are in contemporary continental philosophy, especially the twentieth-century French phenomenological tradition, and the phenomenological investigation of issues in environmental philosophy and the philosophy of nature. He is editor of *Merleau-Ponty: Critical Assessments of Leading Philosophers*, co-editor of *Merleau-Ponty's Reading of Husserl*, and *Eco-Phenomenology: Back to the Earth Itself*, and co-translator of Renaud Barbaras, *The Being of Phenomenon: Merleau-Ponty's Ontology*. He is currently completing a monograph on Merleau-Ponty's philosophy of nature.

Index

Abram, David, 199
agriculture, xv, 186, 189, 195–202, 207–18; Amish, 149, 152; ancient Greek, 177–8; as carbon sequestration, 191; and ethanol production, 191–2; health problems due to, 180; industrial, 195–8, 201; neolithic revolution of, 210; and philosophy, 207; and public interest organizations, 196–9, 202–3; rhizomatic, xvi; Salvadoran, 147; Slow Food, 202; transgenic, 181; weedy, 217. *See also* cultivation, etymology of; eggs, production of; irrigation
Amish, xiv, 147–57
Anaximander, 19
Angermeier, Paul, 32–33
animals: in agroecosystem, 189–90; companion, 86; non-human, 87
anthropocentrism, 83, 85–87, 90, 152–3
Appadurai, Arjun, 119, 120
Aristotle, 19, 20, 63–64, 193n1, 209
Arizona, 167–8
Arnold, T. Clay, xiv
Augustine, 159
autopoiesis, 94–95. *See also* systems: autopoietic

Babbitt, Bruce, 165
Bacon, Francis, 4
Bateson, Gregory, 12
Beauvoir, Simone de, 207–8, 212
Benioff, Marc, 201
Berlin Wall, 124–5
Berry, Wendell, 179, 183, 218

Bickerton, Derek, 3, 10, 13–14
binary logic, 97–98
biocentrism, 84
biodiversity, conservation of, 72–74, 77–79, 199. *See also* conservation biology
bioregionalism, 157
biotechnology, xi, 41, 49–52, 56, 197
Bird, Elizabeth, 95–96
boundaries: arbitrary decisions in drawing, 181–2; and bioregionalism, 157; co-constitution of, xiii, 121–2; human-animal, 7–8; masked, 188–9; nature-culture, 3–7, 21, 31–32, 34, 35, 122, 209, 217–8; spatial, 20–22; taxonomy of, x, xv, 126–7, 179, 185, 191–2; temporal, 21–24, 28–30, 117; transitions between, 113–5, 117–8, 122–3, 126–7. *See also* boundary object
boundary object, 116
Brown, Charles S., xii
Butler, Judith, 120

Callicott, J. Baird, xi, 122
Campbell, Joseph, xv, 213, 214
cancer, 203–4
Carson, Rachel, 121, 178
Cavalli-Sforza, Luigi, 9–10, 13
Clark, Ira, 167
Clements, Frederic, 30
climate change: in Little Ice Age, 30; temporal scale of, 27–28
Colorado, 167–8
community, xiv, 31, 145, 150, 151, 154–9, 165, 170n4; climax, 30;

community, *Cont.*
  irrigation, in American Southwest, 163–4; moral, 83; and place, 150, 158
conservation biology, 32, 36
Copernicus, Nicolaus, 53
coyotes, 62, 66, 77–8
Cronon, William, 217–8
cultivation, etymology of, 208–9

Darwin, Charles, xi, 4, 5, 7, 13, 31, 34, 41, 42, 47–49, 52, 53, 55, 56, 65; on the nature of species, 64
Deacon, Terrence W., 9
death, xvi, 212–5, 216
Deep Ecology movement, 83
Deleuze, Gilles, xiii, xv, 95, 105–8, 213, 215–6
Dempster, Beth, xii–xiii
Descartes, René, xii, 4, 41, 42–44, 47, 52, 53, 55, 84; metaphysical dualism of, 42–43, 85, 196, 200; on substance, 42–43, 44
determinism, Newtonian, 47, 48
deterritorialization, 215
Donnelley, Stachan, xi
Donoghue, Michael, 68
dualism, 119, 216; metaphysical, 42–43, 44; mind-body, 4, 42
Dunbar, Robert, 183

ecofeminism, 84, 207
economic growth, externalization and, 178
economics: and global accounting, 178–9; and middle-ground accounting, 179, 182, 189, 192; neoclassical, 179–80, 194n13
ecosystem, 22, 104–6
ecotone, 120–1
Edwards, Nelta, 203–4
eggs, production of, xv, 183–91
Ehrlich, Paul, 9
El Savador, xiv, 145–7, 149–53, 156, 158; agriculture in, 147
Endangered Species Act, xii, 61, 72–77
environmental crisis, 83, 90

environmental justice: movement, 153; promotion of, 198
Ereshefsky, Marc, 68–69
ethics, 49–52, 55–58; environmental, 35–37
Evernden, Neil, 15n5
evolution, xi, 31, 47, 63; cultural, 10–11, 21, 33–36; Darwinian, xi; human, 8–9, 34, 36; of human brain, 35; Lamarckian, xi, 21, 33–34, 36; neo-Darwinian, 45; temporal scale of, 25, 37
Evolutionarily Significant Units (ESU), 73–74, 75, 76
evolutionary units, 73–74

faith, 211–2, 216
fish, hatchery raised, 51, 76
Foreman, Dave, 217
forests: Douglas fir, 117; management of, 133–40; rain-, Amazonian, 30; rain-, West Coast, 93–94, 101, 104, 105
Foucault, Michel, 120, 127
Freud, Sigmund, 211
Fritzell, Peter, 31–32, 35

Gadamer, Hans-Georg, 5
Garé, Arran, 159
geomorphology, temporal scale of, 25–26
Glover, Jerry, xv
good, social, xiv, 162
Goodall, Jane, 124–5
Grene, Marjorie, 3, 12
Griesemeier, James, 116
Guattari, Félix, xiii, xvi, 95, 105–8, 213, 215–6
Gulf of Mexico, hypoxic zone in, 114, 178–80, 189, 195

Hagedoorn, A. L., 189–90
Haraway, Donna, 14n1, 123
Heidegger, Martin, xv, 15n2, 91n5, 122, 196, 199–201, 204, 212; on agriculture, 208, 211; Being-in-the-world, 200; and environmentalism, 200; on equipment, 201

Heinz, Theodore, 164
Heraclitus, 57
Hesiod, 209
Hill, Johnson, 209
Hirsch, Bruce, xv
Holling, C. S., 22–4, 27
Hume, David, 88
Husserl, Edmund, 84–85, 89

India, forest management and land use practices in, 133–4, 137
institutions, role of in mediating land use practices, 137–40
interdisciplinarity, ix–x
irrigation, 164, 165–6

Jackson, Wes, xv, 157
Jensen, Adolf, 213
Jenson, Jon, xii
Jonas, Hans, 46, 49, 50, 212, 215

Kimmins, James, 104
Kitcher, Philip, 68
Klaver, Irene, xiii
Kohák, Erazim, 210, 212
Kristeva, Julia, 14, 14n2

Lamarck, Jean–Baptiste, 33–34
Land Institute, 191
language, 3, 5, 7–14
Latour, Bruno, 114, 127
Leakey, Richard, 8
Leopold, Aldo, 31, 32, 35, 53, 83, 90, 119, 121, 123, 126, 196
Levine, Daniel, 155
Lichatowich, Jim, 76, 77
Lieberman, Philip, 8
Linnaeus, Carl, 63, 69, 127. *See also* taxonomy, Linnaean
Logsdon, Gene, 190
Los Angeles, 161
love, 129
Lumsden, Charles, 34
Lutten, Dan, 192

Maddock, Tara, 136

Mansfield, Nick, 106
Maturana, Humberto, xiii, 94, 101, 103
Mayr, Ernst, 47, 48, 64
McGinnis, Michael Vincent, 157
McKinney, Matthew, 164
Melville, Herman, 127
Merleau-Ponty, Maurice, 10, 120
Michaels, Anne, 113, 119
Mishler, Brent, 68, 70–71
moral economy, xiv, 161–2, 165–70
Muir, John, 90
music, 48, 54

Nabhan, Gary Paul, 218
Naess, Arne, 90
Nancy, Jean-Luc, 120
narrative, 12–13
National Marine Fisheries Service (NMFS), 75–76
New Mexico, 167
Noss, Reed, 74

Oelschlaeger, Max, x–xi
ontology: agnosticism about, 95–96; dualistic, 42–44, 204; and ethics, 55–58; monistic, 44–45; multiple, 204; realist, 203–4; and social constructivism, 203–4; substance, 42–45. *See also* dualism: metaphysical; relatedness, ontological
organism, 102, 106
Ostfeld, Richard, 36
oxymorons, 118–20, 121

Pavri, Firooza, xiii
Peterson, Anna L., xiv
phenomenology, 84–85, 87, 89
Pickett, Steward, 36
Plato, 20, 78, 193n1
Pleistocene, extinctions during, 177–8
precautionary principle, 180–81
property rights, 137
public interest organizations, 196–9, 202–3

Quinn, Daniel, 210, 212, 217

rationality, 84–85, 90; as defining human characteristic, 20, 32; instrumental, xii, 85, 90; moral, xii, 87, 90
relatedness, ontological, 42–43, 45, 56
religion, xiv, 7, 145–59; Anabaptist, 147–8, 151–5; Catholic, 145–6, 149–156; Protestant, 155; of sacrifice, 213
Rezendes, Paul, 79
rhizome, xiii, 95, 105–8, 215–8; versus tree, 216–7
Robbins, Paul, 136
Rolston, Holmes, III, 150
Rousseau, Jean-Jacques, 209–10, 211, 217
Ruse, Michael, 67

sacrifice, 211–3, 216
salmon, Pacific, 73–77
Sartre, Jean-Paul, 91n5
Schlesinger, William, 191
science, objectivity of, 63
seed, xv, 211–3, 216–7; and temporal orientation, 211–2
Senge, Peter, 204
Shepard, Paul, 211, 212, 217
Sherow, James, 164
Silko, Leslie Marmon, 123–4, 126
Smythe, William, 164
species: and the Biological Species Concept, 64; concept of, xii, 61–79; in conservation efforts, 72–79; as essential types, 48, 64; extinction of, 37, 177–8; and phylogenetics, 64–65; pluralism of, 67–69. *See also* Endangered Species Act
Spengler, Oswald, 214
Spinoza, Baruch, xii, 41, 42, 44–45, 47–49, 52, 53, 55, 56; on "psychophysical parallelism," 45
Star, Susan Leigh, 116
Stoll, David, 155
Stuermann, Walter, 209
sustainability, 13, 94, 116, 149, 151–2; ontological basis for, 204

sympoiesis, 94–95. *See also* systems: sympoietic
systems, 96–97, 105, 192, 204; autopoietic, xiii, 101, 103, 105–6, 108; boundaryless, xii–xiii, 94–95, 97–101, 103–4, 107–8; organizational closure of, 102–3; structural coupling of, 102–3; sympoietic, xii–xiii, 95, 101, 103, 105–8

taxonomy, Linnaean, 69–71, 127
technology: as defining human characteristic, 33; evolution of, 33; remote sensing, and forest management, xiv, 133–6, 138–40; transparent interfaces of, 201–2. *See also* biotechnology
teleology, 47
Texas: birds of, 127–9; "Don't Mess with" slogan, 115–6; El Paso, 167
Thales, 19
Thompson, Paul, 207
Thoreau, Henry David, 90, 127, 196
Toadvine, Ted, xv–xvi
tobacco, health effects of, 180

United Nations, 147

value: intrinsic, 84, 86; natural, 46, 52, 57; pre-reflective experiences of, 87–90
Van Valen, Leigh, 65
Varela, Francisco, xiii, 94, 101, 103
Vergil, 209

Walzer, Michael, 169
water, xiv, 19, 114; commodity view of, 169–70; moral economy of, 161–70; rights, 161, 164, 165–8, 171n10; scarcity in American West, 162–3
watershed partnerships, 164–5
Wayne, Robert, 66
weeds, 217
Weinberg, Gerald, 96
West, Lewis and Clark's discovery of the, 115, 121
Whitehead, Alfred North, 46, 55

whooping crane, 128–9
wilderness: myth of, 217; preservation of, 199
Williams, William Carlos, 129
Wilson, Edward O., 34
Wittgenstein, Ludwig, xiii, 114
wolves, 61–62, 65–66, 75, 77–78, 80n4
worldview, ignorance-based, 182

Wright Act, 165–6

Yellowstone National Park, 61, 78, 80n4, 121–2; Old Faithful, 113
Yoder, John Howard, 154
Young, Brigham, 164

Žižek, Slavoj, 125p